In accordance with Federal civil rights law and U.S. Department of Agriculture (USDA) civil rights regulations and policies, the USDA, its Agencies, offices, and employees, and institutions participating in or administering USDA programs are prohibited from discriminating based on race, color, national origin, religion, sex, gender identity (including gender expression), sexual orientation, disability, age, marital status, family/parental status, income derived from a public assistance program, political beliefs, or reprisal or retaliation for prior civil rights activity, in any program or activity conducted or funded by USDA (not all bases apply to all programs). Remedies and complaint filing deadlines vary by program or incident.

Persons with disabilities who require alternative means of communication for program information (e.g., Braille, large print, audiotape, American Sign Language, etc.) should contact the responsible Agency or USDA's TARGET Center at (202) 720-2600 (voice and TTY) or contact USDA through the Federal Relay Service at (800) 877-8339. Additionally, program information may be made available in languages other than English.

To file a program discrimination complaint, complete the USDA Program Discrimination Complaint Form, AD-3027, found online at How to File a Program Discrimination Complaint (www.ascr.usda.gov/filing-program-discrimination-complaint-usda-customer) and at any USDA office or write a letter addressed to USDA and provide in the letter all of the information requested in the form. To request a copy of the complaint form, call (866) 632-9992. Submit your completed form or letter to USDA by: (1) mail: U.S. Department of Agriculture, Office of the Assistant Secretary for Civil Rights, 1400 Independence Avenue, SW, Washington, D.C. 20250-9410; (2) fax: (202) 690-7442; or (3) email: program.intake@usda.gov.

USDA is an equal opportunity provider, employer, and lender.

Issued January 2017

TABLE OF CONTENTS

BLUE BOOK

Table of Contents

BLUE BOOK

Table of Contents

PART 1 – DEFINITION OF TERMS

Section   *Page*

PART 2 – REGULATIONS

Section   *Page*

Subpart A – Licensing

Subpart B – Registration

Subpart C – Research Facilities

Subpart D – Attending Veterinarian and Adequate Veterinary Care

Subpart E – Identification of Animals

Subpart F – Stolen Animals

Subpart G – Records

Subpart H – Compliance With Standards and Holding Period

Subpart I – Miscellaneous

Subpart J—Importation of Live Dogs

PART 3 – STANDARDS

Section   *Page*

**Subpart A – Specifications for the Humane Handling,
          Care, Treatment, and Transportation of Dogs and Cats**

**Subpart B – Specifications for the Humane Handling,
Care, Treatment, and Transportation of Guinea Pigs
and Hamsters**

Subpart C – Specifications for the Humane Handling, Care, Treatment and Transportation of Rabbits

BLUE BOOK
Table of Contents

**Subpart F – Specifications for the Humane Handling,
         Care, Treatment, and Transportation of Warmblooded
         Animals Other Than Dogs, Cats, Rabbits, Hamsters,
         Guinea Pigs, Nonhuman Primates, and Marine Mammals**

**PART 4 – RULES OF PRACTICE GOVERNING
PROCEEDINGS UNDER THE ANIMAL WELFARE ACT**

Introduction

Passed by Congress in 1966, the Animal Welfare Act (AWA) sets general standards for humane care and treatment that must be provided for certain animals that are bred for commercial sale, sold sight unseen (Internet sales), exhibited to the public, used in biomedical research, or transported commercially. Congress assigned the U.S. Department of Agriculture (USDA) the responsibility for enforcing the AWA. The Animal and Plant Health Inspection Service (APHIS) is the agency within USDA responsible for ensuring this occurs.

APHIS has published the *Animal Welfare Act and Animal Welfare Regulations,* known as the "Blue Book," as a tool to improve compliance among our licensees and registrants and to enhance the consistency of inspections by our field inspectors. The Blue Book consolidates into one source the AWA and the applicable regulations and standards.

Text was pulled directly from two official sources: the **Animal Welfare Act** [*United States Code*, Title 7 (Agriculture), Chapter 54 (Transportation, Sale, and Handling of Certain Animals), Sections 2131– 2159]; and the **Animal Welfare Regulations** [*Code of Federal Regulations*, Title 9 (Animals and Animal Products), Chapter 1 (Animal and Plant Health Inspection Service, Department of Agriculture), Subchapter A (Animal Welfare), Parts 1–4]. Every effort has been made to ensure that this book is accurate.

The AWA is a Federal law; the associated regulations interpret the law into enforceable standards. USDA can amend existing regulations and/or create new regulations, but only Congress can change the actual law. Prior to altering the regulations in any way, APHIS publicizes the proposed rule and encourages interested parties to provide comments. All comments are reviewed before determining a final course of action.

CONTROL DATA Control data is located at the top and bottom of each page to help readers keep track of where they are and to be aware of updates to specific chapters, appendixes, and other sections.

FOOTNOTES The footnotes include text, figures, and tables. Text footnotes are located at the bottom of a page. When space allows, footnotes involving figures and tables are located directly below the associated figure or table. Otherwise, the footnotes are placed as close to the figure or table as possible.

UPDATES APHIS maintains an electronic copy of the Blue Book[3] on our agency's Animal Care Web site: **www.aphis.usda.gov /animal-welfare**. The electronic version contains the most up-to-date information.

When the Blue Book is revised, Animal Care notifies interested parties via the USDA-APHIS Stakeholder Registry. To subscribe, please register here: **https://public.govdelivery.com/accounts/USDAAPHIS/subscriber/new**. You can select the individual topics that interest you, or you can choose to receive all of our agency's messages.

ORDERING PRINTED COPIES APHIS encourages you to use the electronic version of the Blue Book, as it will always be the most current, but you can still order printed copies. The link to the APHIS publications ordering page is: **www.aphis.usda.gov/publications**.

To order, click "PLACE YOUR ORDER HERE," and then choose "ANIMAL WELFARE" in the dropdown box. Click "Search This View," and then type "Animal Welfare Act and Animal Welfare Regulations" in the Keyword box. Click "Perform Search," and then click on the AC-AWA&R link. Type the quantity you are requesting, and then click "Add to Cart."

When ready, click "View Cart," and fill out the contact and shipping information on the order form. Once all is completed, click "Submit Order."

If you do not have access to the Internet, you may call APHIS' Publication and Distribution Management office to order copies at (301) 851-2633.

QUESTIONS OR CONCERNS If you have questions or concerns regarding the Blue Book, please contact Animal Care's National Policy staff at (301) 851-3751.

Animal Welfare Act
As of Jan. 1, 2017

As found in the United States Code

Title 7 – Agriculture
Chapter 54 – Transportation, Sale,
and Handling of Certain Animals

Sections 2131 - 2159

4

ANIMAL WELFARE ACT

§ 2131 - Congressional statement of policy

The Congress finds that animals and activities which are regulated under this chapter are either in interstate or foreign commerce or substantially affect such commerce or the free flow thereof, and that regulation of animals and activities as provided in this chapter is necessary to prevent and eliminate burdens upon such commerce and to effectively regulate such commerce, in order –

(1) to insure that animals intended for use in research facilities or for exhibition purposes or for use as pets are provided humane care and treatment;

(2) to assure the humane treatment of animals during transportation in commerce; and

(3) to protect the owners of animals from the theft of their animals by preventing the sale or use of animals which have been stolen.

The Congress further finds that it is essential to regulate, as provided in this chapter, the transportation, purchase, sale, housing, care, handling, and treatment of animals by carriers or by persons or organizations engaged in using them for research or experimental purposes or for exhibition purposes or holding them for sale as pets or for any such purpose or use.

For a full history of all of the Animal Welfare Act amendments, please visit: http://www.aphis.usda.gov/animal_welfare/downloads/awa/awa.pdf

§ 2132 - Definitions

In this chapter:

(a) The term "**person**" includes any individual, partnership, firm, joint stock company, corporation, association, trust, estate, or other legal entity.

(b) The term "**Secretary**" means the Secretary of Agriculture of the United States or his representative who shall be an employee of the United States Department of Agriculture.

(c) The term "**commerce**" means trade, traffic, transportation, or other commerce –

(1) between a place in a State and any place outside of such State, or between points within the same State but through any place outside thereof, or within any territory, possession, or the District of Columbia;

(2) which affects trade, traffic, transportation, or other commerce described in paragraph **(1)**.

(d) The term "**State**" means a State of the United States, the District of Columbia, the Commonwealth of Puerto Rico, the Virgin Islands, Guam, American Samoa, or any other territory or possession of the United States.

(e) The term "**research facility**" means any school (except an elementary or secondary school), institution, organization, or person that uses or intends to use live animals in research, tests, or experiments, and that **(1)** purchases or transports live animals in commerce, or **(2)** receives funds under a grant, award, loan, or contract from a department, agency, or instrumentality of the United States for the purpose of carrying out research, tests, or experiments: *Provided,* That the Secretary may exempt, by regulation, any such school, institution, organization, or person that does not use or intend to use live dogs or cats, except those schools, institutions, organizations, or persons, which use substantial numbers (as determined by the Secretary) of live animals the principal function of which schools, institutions, organizations, or persons, is biomedical research or testing, when in the judgment of the Secretary, any such exemption does not vitiate the purpose of this chapter.

(f) The term "**dealer**" means any person who, in commerce, for compensation or profit, delivers for transportation, or transports, except as a carrier, buys, or sells, or negotiates the purchase or sale of, **(1)** any dog or other animal whether alive or dead for research, teaching, exhibition, or use as a pet, or **(2)** any dog for hunting, security, or breeding purposes. Such term does not include a retail pet store (other than a retail pet store which sells any animals to a research facility, an exhibitor, or another dealer).

(g) The term "**animal**" means any live or dead dog, cat, monkey (nonhuman primate mammal), guinea pig, hamster, rabbit, or such other warm-blooded animal, as the Secretary may determine is being used, or is intended for use, for research, testing, experimentation, or exhibition purposes, or as a pet; but such term excludes **(1)** birds, rats of the genus Rattus, and mice of the genus Mus, bred for use in research, **(2)** horses not used for research purposes, and **(3)** other farm animals, such as, but not limited to livestock or poultry, used or intended for use as food or fiber, or livestock or poultry used or intended for use for improving animal nutrition, breeding, management, or production efficiency, or for improving the quality of food or fiber. With respect to a dog, the term means all dogs including those used for hunting, security, or breeding purposes.

(h) The term "**exhibitor**" means any person (public or private) exhibiting any animals, which were purchased in commerce or the intended distribution of which affects commerce, or will affect commerce, to the public for compensation, as determined by the Secretary, and such term includes carnivals, circuses, and zoos exhibiting such animals whether operated for profit or not; but such term excludes retail pet stores, an owner of a common, domesticated household pet who derives less than a substantial portion

6

of income from a nonprimary source (as determined by the Secretary) for exhibiting an animal that exclusively resides at the residence of the pet owner, organizations sponsoring and all persons participating in State and country fairs, livestock shows, rodeos, purebred dog and cat shows, and any other fairs or exhibitions intended to advance agricultural arts and sciences, as may be determined by the Secretary.

(i) The term "**intermediate handler**" means any person including a department, agency, or instrumentality of the United States or of any State or local government (other than a dealer, research facility, exhibitor, any person excluded from the definition of a dealer, research facility, or exhibitor, an operator of an auction sale, or a carrier) who is engaged in any business in which he receives custody of animals in connection with their transportation in commerce.

(j) The term "**carrier**" means the operator of any airline, railroad, motor carrier, shipping line, or other enterprise, which is engaged in the business of transporting any animals for hire.

(k) The term "**Federal agency**" means an Executive agency as such term is defined in section 105 of title 5, and with respect to any research facility means the agency from which the research facility receives a Federal award for the conduct of research, experimentation, or testing, involving the use of animals.

(l) The term "**Federal award for the conduct of research, experimentation, or testing, involving the use of animals**" means any mechanism (including a grant, award, loan, contract, or cooperative agreement) under which Federal funds are provided to support the conduct of such research.

(m) The term "**quorum**" means a majority of the Committee members.

(n) The term "**Committee**" means the Institutional Animal Committee established under section 2143(b) of this title.

(o) The term "**Federal research facility**" means each department, agency, or instrumentality of the United States which uses live animals for research or experimentation.

§ 2133 - Licensing of dealers and exhibitors

The Secretary shall issue licenses to dealers and exhibitors upon application therefor in such form and manner as he may prescribe and upon payment of such fee established pursuant to 2153 of this title: *Provided,* That no such license shall be issued until the dealer or exhibitor shall have demonstrated that his facilities comply with the standards promulgated by the Secretary pursuant to section 2143 of this title: *Provided, however,* That a dealer or exhibitor shall not be required to obtain a license as a dealer or exhibitor under this chapter if the size of the business is determined by the

Secretary to be de minimis. The Secretary is further authorized to license, as dealers or exhibitors, persons who do not qualify as dealers or exhibitors within the meaning of this chapter upon such persons' complying with the requirements specified above and agreeing, in writing, to comply with all the requirements of this chapter and the regulations promulgated by the Secretary hereunder.

§ 2134 - Valid license for dealers and exhibitors required

No dealer or exhibitor shall sell or offer to sell or transport or offer for transportation, in commerce, to any research facility or for exhibition or for use as a pet any animal, or buy, sell, offer to buy or sell, transport or offer for transportation, in commerce, to or from another dealer or exhibitor under this chapter any animals, unless and until such dealer or exhibitor shall have obtained a license from the Secretary and such license shall not have been suspended or revoked.

§ 2135 - Time period for disposal of dogs or cats by dealers or exhibitors

No dealer or exhibitor shall sell or otherwise dispose of any dog or cat within a period of five business days after the acquisition of such animal or within such other period as may be specified by the Secretary: *Provided,* That operators of auction sales subject to section 2142 of this title shall not be required to comply with the provisions of this section.

§ 2136 - Registration of research facilities, handlers, carriers and unlicensed exhibitors

Every research facility, every intermediate handler, every carrier, and every exhibitor not licensed under section 2133 of this title shall register with the Secretary in accordance with such rules and regulations as he may prescribe.

§ 2137 - Purchase of dogs or cats by research facilities prohibited except from authorized operators of auction sales and licensed dealers or exhibitors

It shall be unlawful for any research facility to purchase any dog or cat from any person except an operator of an auction sale subject to section 2142 of this title or a person holding a valid license as a dealer or exhibitor issued by the Secretary pursuant to this chapter unless such person is exempted from obtaining such license under section 2133 of this title.

§ 2138 - Purchase of dogs or cats by United States Government facilities prohibited except from authorized operators of auction sales and licensed dealers or exhibitors

No department, agency, or instrumentality of the United States which uses animals for research or experimentation or exhibition shall purchase or otherwise acquire any dog or cat for such purposes from any person except an operator of an auction sale subject to section 2142 of this title or a person holding a valid license as a dealer or exhibitor issued by the Secretary pursuant to this chapter unless such person is exempted from obtaining such license under section 2133 of this title.

§ 2139 - Principal-agent relationship established

When construing or enforcing the provisions of this chapter, the act, omission, or failure of any person acting for or employed by a research facility, a dealer, or an exhibitor or a person licensed as a dealer or an exhibitor pursuant to the second sentence of section 2133 of this title, or an operator of an auction sale subject to section 2142 of this title, or an intermediate handler, or a carrier, within the scope of his employment or office, shall be deemed the act, omission, or failure of such research facility, dealer, exhibitor, licensee, operator of an auction sale, intermediate handler, or carrier, as well as of such person.

§ 2140 - Recordkeeping by dealers, exhibitors, research facilities, intermediate handlers, and carriers

Dealers and exhibitors shall make and retain for such reasonable period of time as the Secretary may prescribe, such records with respect to the purchase, sale, transportation, identification, and previous ownership of animals as the Secretary may prescribe. Research facilities shall make and retain such records only with respect to the purchase, sale, transportation, identification, and previous ownership of live dogs and cats. At the request of the Secretary, any regulatory agency of the Federal Government which requires records to be maintained by intermediate handlers and carriers with respect to the transportation, receiving, handling, and delivery of animals on forms prescribed by the agency, shall require there to be included in such forms, and intermediate handlers and carriers shall include in such forms, such information as the Secretary may require for the effective administration of this chapter. Such information shall be retained for such reasonable period of time as the Secretary may prescribe. If regulatory agencies of the Federal Government do not prescribe requirements for any such forms, intermediate handlers and carriers shall make and retain for such reasonable period as the Secretary may prescribe such records with respect to the transportation, receiving, handling, and delivery of animals as the Secretary may prescribe.

Such records shall be made available at all reasonable times for inspection and copying by the Secretary.

§ 2141 - Marking and identification of animals

All animals delivered for transportation, transported, purchased, or sold, in commerce, by a dealer or exhibitor shall be marked or identified at such time and in such humane manner as the Secretary may prescribe: *Provided,* That only live dogs and cats need be so marked or identified by a research facility.

§ 2142 - Humane standards and recordkeeping requirements at auction sales

The Secretary is authorized to promulgate humane standards and recordkeeping requirements governing the purchase, handling, or sale of animals, in commerce, by dealers, research facilities, and exhibitors at auction sales and by the operators of such auction sales. The Secretary is also authorized to require the licensing of operators of auction sales where any dogs or cats are sold, in commerce, under such conditions as he may prescribe, and upon payment of such fee as prescribed by the Secretary under section 2153 of this title.

§ 2143 - Standards and certification process for humane handling, care, treatment, and transportation of animals

(a) Promulgation of standards, rules, regulations, and orders; requirements; research facilities; State authority

(1) The Secretary shall promulgate standards to govern the humane handling, care, treatment, and transportation of animals by dealers, research facilities, and exhibitors.

(2) The standards described in paragraph **(1)** shall include minimum requirements –

(A) for handling, housing, feeding, watering, sanitation, ventilation, shelter from extremes of weather and temperatures, adequate veterinary care, and separation by species where the Secretary finds necessary for humane handling, care, or treatment of animals; and

(B) for exercise of dogs, as determined by an attending veterinarian in accordance with general standards promulgated by the Secretary, and for a physical environment adequate to promote the psychological well-being of primates.

(3) In addition to the requirements under paragraph (2), the standards described in paragraph (1) shall, with respect to animals in research facilities, include requirements –

(A) for animal care, treatment, and practices in experimental procedures to ensure that animal pain and distress are minimized, including

10

adequate veterinary care with the appropriate use of anesthetic, analgesic, tranquilizing drugs, or euthanasia;

(B) that the principal investigator considers alternatives to any procedure likely to produce pain to or distress in an experimental animal;

(C) in any practice which could cause pain to animals –

(i) that a doctor of veterinary medicine is consulted in the planning of such procedures;

(ii) for the use of tranquilizers, analgesics, and anesthetics;

(iii) for pre-surgical and post-surgical care by laboratory workers, in accordance with established veterinary medical and nursing procedures;

(iv) against the use of paralytics without anesthesia; and

(v) that the withholding of tranquilizers, anesthesia, analgesia, or euthanasia when scientifically necessary shall continue for only the necessary period of time;

(D) that no animal is used in more than one major operative experiment from which it is allowed to recover except in cases of –

(i) scientific necessity; or

(ii) other special circumstances as determined by the Secretary; and

(E) that exceptions to such standards may be made only when specified by research protocol and that any such exception shall be detailed and explained in a report outlined under paragraph (7) and filed with the Institutional Animal Committee.

(4) The Secretary shall also promulgate standards to govern the transportation in commerce, and the handling, care, and treatment in connection therewith, by intermediate handlers, air carriers, or other carriers, of animals consigned by any dealer, research facility, exhibitor, operator of an auction sale, or other person, or any department, agency, or instrumentality of the United States or of any State or local government, for transportation in commerce. The Secretary shall have authority to promulgate such rules and regulations as he determines necessary to assure humane treatment of animals in the course of their transportation in commerce including requirements such as those with respect to containers, feed, water, rest, ventilation, temperature, and handling.

(5) In promulgating and enforcing standards established pursuant to this section, the Secretary is authorized and directed to consult experts, including outside consultants where indicated.

(6)(A) Nothing in this chapter –

(i) except as provided in paragraphs[1] (7) of this subsection, shall be construed as authorizing the Secretary to promulgate rules, regulations, or

---

1 So in original. Probably should be "paragraph".

orders with regard to the design, outlines, or guidelines of actual research or experimentation by a research facility as determined by such research facility;

(ii) except as provided[2] subparagraphs (A) and (C)(ii) through (v) of paragraph (3) and paragraph (7) of this subsection, shall be construed as authorizing the Secretary to promulgate rules, regulations, or orders with regard to the performance of actual research or experimentation by a research facility as determined by such research facility; and

(iii) shall authorize the Secretary, during inspection, to interrupt the conduct of actual research or experimentation.

(B) No rule, regulation, order, or part of this chapter shall be construed to require a research facility to disclose publicly or to the Institutional Animal Committee during its inspection, trade secrets or commercial or financial information which is privileged or confidential.

(7)(A) The Secretary shall require each research facility to show upon inspection, and to report at least annually, that the provisions of this chapter are being followed and that professionally acceptable standards governing the care, treatment, and use of animals are being followed by the research facility during actual research or experimentation.

(B) In complying with subparagraph (A), such research facilities shall provide –

(i) information on procedures likely to produce pain or distress in any animal and assurances demonstrating that the principal investigator considered alternatives to those procedures;

(ii) assurances satisfactory to the Secretary that such facility is adhering to the standards described in this section; and

(iii) an explanation for any deviation from the standards promulgated under this section.

(8) Paragraph (1) shall not prohibit any State (or a political subdivision of such State) from promulgating standards in addition to those standards promulgated by the Secretary under paragraph (1).

(b) Research facility Committee; establishment, membership, functions, etc.

(1) The Secretary shall require that each research facility establish at least one Committee. Each Committee shall be appointed by the chief executive officer of each such research facility and shall be composed of not fewer than three members. Such members shall possess sufficient ability to assess animal care, treatment, and practices in experimental research as determined by the needs of the research facility and shall represent society's concerns regarding the welfare of animal subjects used at such facility. Of the members of the Committee –

---

2 *So in original. Probably should be followed by "in".*

(**A**) at least one member shall be a doctor of veterinary medicine;

(**B**) at least one member –

(**i**) shall not be affiliated in any way with such facility other than as a member of the Committee;

(**ii**) shall not be a member of the immediate family of a person who is affiliated with such facility; and

(**iii**) is intended to provide representation for general community interests in the proper care and treatment of animals; and

(**C**) in those cases where the Committee consists of more than three members, not more than three members shall be from the same administrative unit of such facility.

(**2**) A quorum shall be required for all formal actions of the Committee, including inspections under paragraph (3).

(**3**) The Committee shall inspect at least semiannually all animal study areas and animal facilities of such research facility and review as part of the inspection –

(**A**) practices involving pain to animals, and

(**B**) the condition of animals, to ensure compliance with the provisions of this chapter to minimize pain and distress to animals. Exceptions to the requirement of inspection of such study areas may be made by the Secretary if animals are studied in their natural environment and the study area is prohibitive to easy access.

(**4**)(**A**) The Committee shall file an inspection certification report of each inspection at the research facility. Such report shall –

(**i**) be signed by a majority of the Committee members involved in the inspection;

(**ii**) include reports of any violation of the standards promulgated, or assurances required, by the Secretary, including any deficient conditions of animal care or treatment, any deviations of research practices from originally approved proposals that adversely affect animal welfare, any notification to the facility regarding such conditions, and any corrections made thereafter;

(**iii**) include any minority views of the Committee; and

(**iv**) include any other information pertinent to the activities of the Committee.

(**B**) Such report shall remain on file for at least three years at the research facility and shall be available for inspection by the Animal and Plant Health Inspection Service and any funding Federal agency.

(**C**) In order to give the research facility an opportunity to correct any deficiencies or deviations discovered by reason of paragraph (3), the Committee shall notify the administrative representative of the research facility of any deficiencies or deviations from the provisions of this chapter. If, after notification and an opportunity for correction, such deficiencies or deviations remain uncorrected, the Committee shall notify (in writing) the Animal and Plant Health Inspection Service and the funding Federal agency of such deficiencies or deviations.

(5) The inspection results shall be available to Department of Agriculture inspectors for review during inspections. Department of Agriculture inspectors shall forward any Committee inspection records which include reports of uncorrected deficiencies or deviations to the Animal and Plant Health Inspection Service and any funding Federal agency of the project with respect to which such uncorrected deficiencies and deviations occurred.

(c) Federal research facilities; establishment, composition, and responsibilities of Federal Committee

In the case of Federal research facilities, a Federal Committee shall be established and shall have the same composition and responsibilities provided in subsection (b) of this section, except that the Federal Committee shall report deficiencies or deviations to the head of the Federal agency conducting the research rather than to the Animal and Plant Health Inspection Service. The head of the Federal agency conducting the research shall be responsible for –

(1) all corrective action to be taken at the facility; and

(2) the granting of all exceptions to inspection protocol.

(d) Training of scientists, animal technicians, and other personnel involved with animal care and treatment at research facilities

Each research facility shall provide for the training of scientists, animal technicians, and other personnel involved with animal care and treatment in such facility as required by the Secretary. Such training shall include instruction on –

(1) the humane practice of animal maintenance and experimentation;

(2) research or testing methods that minimize or eliminate the use of animals or limit animal pain or distress;

(3) utilization of the information service at the National Agricultural Library, established under subsection (e) of this section; and

(4) methods whereby deficiencies in animal care and treatment should be reported.

(e) Establishment of information service at National Agricultural Library; service functions

The Secretary shall establish an information service at the National Agricultural Library. Such service shall, in cooperation with the National Library of Medicine, provide information –

(1) pertinent to employee training;

(2) which could prevent unintended duplication of animal experimentation as determined by the needs of the research facility; and

14

(3) on improved methods of animal experimentation, including methods which could –

   (A) reduce or replace animal use; and

   (B) minimize pain and distress to animals, such as anesthetic and analgesic procedures.

(f) [3] **Suspension or revocation of Federal support for research projects; prerequisites; appeal procedure**

In any case in which a Federal agency funding a research project determines that conditions of animal care, treatment, or practice in a particular project have not been in compliance with standards promulgated under this chapter, despite notification by the Secretary or such Federal agency to the research facility and an opportunity for correction, such agency shall suspend or revoke Federal support for the project. Any research facility losing Federal support as a result of actions taken under the preceding sentence shall have the right of appeal as provided in sections 701 through 706 of title 5.

(f) [3] **Veterinary certificate; contents; exceptions**

No dogs or cats, or additional kinds or classes of animals designated by regulation of the Secretary, shall be delivered by any dealer, research facility, exhibitor, operator of an auction sale, or department, agency, or instrumentality of the United States or of any State or local government, to any intermediate handler or carrier for transportation in commerce, or received by any such handler or carrier for such transportation from any such person, department, agency, or instrumentality, unless the animal is accompanied by a certificate issued by a veterinarian licensed to practice veterinary medicine, certifying that he inspected the animal on a specified date, which shall not be more than ten days before such delivery, and, when so inspected, the animal appeared free of any infectious disease or physical abnormality which would endanger the animal or animals or other animals or endanger public health: *Provided*, however, That the Secretary may by regulation provide exceptions to this certification requirement, under such conditions as he may prescribe in the regulations, for animals shipped to research facilities for purposes of research, testing or experimentation requiring animals not eligible for such certification. Such certificates received by the intermediate handlers and the carriers shall be retained by them, as provided by regulations of the Secretary, in accordance with section 2140 of this title.

(g) Age of animals delivered to registered research facilities; power of Secretary to designate additional classes of animals and age limits

No dogs or cats, or additional kinds or classes of animals designated by regulation of the Secretary, shall be delivered by any person to any intermediate handler or carrier for transportation in commerce except to

---

3 So in original. Two subsecs. (f) have been enacted.

registered research facilities if they are less than such age as the Secretary may by regulation prescribe. The Secretary shall designate additional kinds and classes of animals and may prescribe different ages for particular kinds or classes of dogs, cats, or designated animals, for the purposes of this section, when he determines that such action is necessary or adequate to assure their humane treatment in connection with their transportation in commerce.

(h) Prohibition of C.O.D. arrangements for transportation of animals in commerce; exceptions

No intermediate handler or carrier involved in the transportation of any animal in commerce shall participate in any arrangement or engage in any practice under which the cost of such animal or the cost of the transportation of such animal is to be paid and collected upon delivery of the animal to the consignee, unless the consignor guarantees in writing the payment of transportation charges for any animal not claimed within a period of 48 hours after notice to the consignee of arrival of the animal, including, where necessary, both the return transportation charges and an amount sufficient to reimburse the carrier for all out-of-pocket expenses incurred for the care, feeding, and storage of such animals.

§ 2144 - Humane standards for animals by United States Government facilities

Any department, agency, or instrumentality of the United States having laboratory animal facilities shall comply with the standards and other requirements promulgated by the Secretary under sections[1] 2143(a),(f),(g) and (h) of this title. Any department, agency, or instrumentality of the United States exhibiting animals shall comply with the standards promulgated by the Secretary under sections 2143(a), (f), (g), and (h) of this title.

§ 2145 - Consultation and cooperation with Federal, State, and local governmental bodies by Secretary of Agriculture

(a) The Secretary shall consult and cooperate with other Federal departments, agencies, or instrumentalities concerned with the welfare of animals used for research, experimentation or exhibition, or administration of statutes regulating the transportation in commerce or handling in connection therewith of any animals when establishing standards pursuant to section 2143 of this title and in carrying out the purposes of this chapter. The Secretary shall consult with the Secretary of Health and Human Services prior to issuance of regulations. Before promulgating any standard governing the air transportation and handling in connection therewith, of animals, the Secretary shall consult with the Secretary of Transportation who shall have the authority to disapprove any such standard if he notifies the Secretary,

---

1 So in original. Probably should be "section".

within 30 days after such consultation, that changes in its provisions are necessary in the interest of flight safety. The Surface Transportation Board, the Secretary of Transportation, and the Federal Maritime Commission, to the extent of their respective lawful authorities, shall take such action as is appropriate to implement any standard established by the Secretary with respect to a person subject to regulation by it.

(b) The Secretary is authorized to cooperate with the officials of the various States or political subdivisions thereof in carrying out the purposes of this chapter and of any State, local, or municipal legislation or ordinance on the same subject.

§ 2146 - Administration and enforcement by Secretary
(a) Investigations and inspections

The Secretary shall make such investigations or inspections as he deems necessary to determine whether any dealer, exhibitor, intermediate handler, carrier, research facility, or operator of an auction sale subject to section 2142 of this title, has violated or is violating any provision of this chapter or any regulation or standard issued thereunder, and for such purposes, the Secretary shall, at all reasonable times, have access to the places of business and the facilities, animals, and those records required to be kept pursuant to section 2140 of this title of any such dealer, exhibitor, intermediate handler, carrier, research facility, or operator of an auction sale. The Secretary shall inspect each research facility at least once each year and, in the case of deficiencies or deviations from the standards promulgated under this chapter, shall conduct such follow-up inspections as may be necessary until all deficiencies or deviations from such standards are corrected. The Secretary shall promulgate such rules and regulations as he deems necessary to permit inspectors to confiscate or destroy in a humane manner any animal found to be suffering as a result of a failure to comply with any provision of this chapter or any regulation or standard issued thereunder if (1) such animal is held by a dealer, (2) such animal is held by an exhibitor, (3) such animal is held by a research facility and is no longer required by such research facility to carry out the research, test, or experiment for which such animal has been utilized, (4) such animal is held by an operator of an auction sale, or (5) such animal is held by an intermediate handler or a carrier.

(b) Penalties for interfering with official duties

Any person who forcibly assaults, resists, opposes, impedes, intimidates, or interferes with any person while engaged in or on account of the performance of his official duties under this chapter shall be fined not more than $5,000, or imprisoned not more than three years, or both. Whoever, in the commission of such acts, uses a deadly or dangerous weapon shall be fined not more than $10,000, or imprisoned not more than ten years, or both. Whoever kills any person while engaged in or on account of the performance of his official duties under this chapter shall be punished as provided under sections 1111 and 1114 of title 18.

(c) Procedures

For the efficient administration and enforcement of this chapter and the regulations and standards promulgated under this chapter, the provisions (including penalties) of sections 46, 48, 49 and 50 of title 15 (except paragraph (c) through (h) of section 46 and the last paragraph of section 49 of title 15), and the provisions of Title II of the Organized Crime Control Act of 1970, are made applicable to the jurisdiction, powers, and duties of the Secretary in administering and enforcing the provisions of this chapter and to any person, firm, or corporation with respect to whom such authority is exercised. The Secretary may prosecute any inquiry necessary to his duties under this chapter in any part of the United States, including any territory, or possession thereof, the District of Columbia, or the Commonwealth of Puerto Rico. The powers conferred by said sections 49 and 50 of title 15 on the district courts of the United States may be exercised for the purposes of this chapter by any district court of the United States. The United States district courts, the District Court of Guam, the District Court of the Virgin Islands, the highest court of American Samoa, and the United States courts of the other territories, are vested with jurisdiction specifically to enforce, and to prevent and restrain violations of this chapter, and shall have jurisdiction in all other kinds of cases arising under this chapter, except as provided in section 2149(c) of this title.

§ 2147 - Inspection by legally constituted law enforcement agencies

The Secretary shall promulgate rules and regulations requiring dealers, exhibitors, research facilities, and operators of auction sales subject to section 2142 of this title to permit inspection of their animals and records at reasonable hours upon request by legally constituted law enforcement agencies in search of lost animals.

§ 2148 - Importation of live dogs
(a) Definitions
In this section:
(1) Importer
The term "**importer**" means any person who, for purposes of resale, transports into the United States puppies from a foreign country.
(2) Resale
The term "**resale**" includes any transfer of ownership or control of an imported dog of less than 6 months of age to another person, for more than de minimis consideration.

(b) Requirements

(1) In general

Except as provided in paragraph (2), no person shall import a dog into the United States for purposes of resale unless, as determined by the Secretary, the dog –

    (A) is in good health;

    (B) has received all necessary vaccinations; and

    (C) is at least 6 months of age, if imported for resale.

(2) Exception

    (A) In general

The Secretary, by regulation, shall provide an exception to any requirement under paragraph (1) in any case in which a dog is imported for –

        (i) research purposes; or

        (ii) veterinary treatment.

    (B) Lawful importation into Hawaii

Paragraph (1)(C) shall not apply to the lawful importation of a dog into the State of Hawaii from the British Isles, Australia, Guam, or New Zealand in compliance with the applicable regulations of the State of Hawaii and the other requirements of this section, if the dog is not transported out of the State of Hawaii for purposes of resale at less than 6 months of age.

(c) Implementation and regulations

The Secretary, the Secretary of Health and Human Services, the Secretary of Commerce, and the Secretary of Homeland Security shall promulgate such regulations as the Secretaries determine to be necessary to implement and enforce this section.

(d) Enforcement

An importer that fails to comply with this section shall –

    (1) be subject to penalties under section 2149 of this title; and

    (2) provide for the care (including appropriate veterinary care), forfeiture, and adoption of each applicable dog, at the expense of the importer.

§ 2149 - Violations by licensees

(a) Temporary license suspension; notice and hearing; revocation

If the Secretary has reason to believe that any person licensed as a dealer, exhibitor, or operator of an auction sale subject to section 2142 of this title, has violated or is violating any provision of this chapter, or any of the rules or regulations or standards promulgated by the Secretary hereunder, he may suspend such person's license temporarily, but not to exceed 21 days, and after notice and opportunity for hearing, may suspend for such additional period as he may specify, or revoke such license, if such violation is determined to have occurred.

(b) Civil penalties for violation of any section, etc.; separate offenses; notice and hearing; appeal; considerations in assessing penalty; compromise of penalty; civil action by Attorney General for failure to pay penalty; district court jurisdiction; failure to obey cease and desist order

Any dealer, exhibitor, research facility, intermediate handler, carrier, or operator of an auction sale subject to section 2142 of this title, that violates any provision of this chapter, or any rule, regulation, or standard promulgated by the Secretary thereunder, may be assessed a civil penalty by the Secretary of not more than $10,000 for each such violation, and the Secretary may also make an order that such person shall cease and desist from continuing such violation. Each violation and each day during which a violation continues shall be a separate offense. No penalty shall be assessed or cease and desist order issued unless such person is given notice and opportunity for a hearing with respect to the alleged violation, and the order of the Secretary assessing a penalty and making a cease and desist order shall be final and conclusive unless the affected person files an appeal from the Secretary's order with the appropriate United States Court of Appeals. The Secretary shall give due consideration to the appropriateness of the penalty with respect to the size of the business of the person involved, the gravity of the violation, the person's good faith, and the history of previous violations. Any such civil penalty may be compromised by the Secretary. Upon any failure to pay the penalty assessed by a final order under this section, the Secretary shall request the Attorney General to institute a civil action in a district court of the United States or other United States court for any district in which such person is found or resides or transacts business, to collect the penalty, and such court shall have jurisdiction to hear and decide any such action. Any person who knowingly fails to obey a cease and desist order made by the Secretary under this section shall be subject to a civil penalty of $1,500 for each offense, and each day during which such failure continues shall be deemed a separate offense.

(c) Appeal of final order by aggrieved person; limitations; exclusive jurisdiction of United States Courts of Appeals

Any dealer, exhibitor, research facility, intermediate handler, carrier, or operator of an auction sale subject to section 2142 of this title, aggrieved by a final order of the Secretary issued pursuant to this section may, within 60 days after entry of such an order, seek review of such order in the appropriate United States Court of Appeals in accordance with the provisions of sections 2341, 2343 through 2350 of title 28, and such court shall have exclusive jurisdiction to enjoin, set aside, suspend (in whole or in part), or to determine the validity of the Secretary's order.

20

(d) Criminal penalties for violation; initial prosecution brought before United States magistrate judges; conduct of prosecution by attorneys of United States Department of Agriculture

Any dealer, exhibitor, or operator of an auction sale subject to section 2142 of this title, who knowingly violates any provision of this chapter shall, on conviction thereof, be subject to imprisonment for not more than 1 year, or a fine of not more than $2,500, or both. Prosecution of such violations shall, to the maximum extent practicable, be brought initially before United States magistrate judges as provided in section 636 of title 28, and sections 3401 and 3402 of title 18, and, with the consent of the Attorney General, may be conducted, at both trial and upon appeal to district court, by attorneys of the United States Department of Agriculture.

§ 2150 - Repealed

§ 2151 - Rules and regulations

The Secretary is authorized to promulgate such rules, regulations, and orders as he may deem necessary in order to effectuate the purposes of this chapter.

§ 2152 - Separability

If any provision of this chapter or the application of any such provision to any person or circumstances shall be held invalid, the remainder of this chapter and the application of any such provision to persons or circumstances other than those as to which it is held invalid shall not be affected thereby.

§ 2153 - Fees and authorization of appropriations

The Secretary shall charge, assess, and cause to be collected reasonable fees for licenses issued. Such fees shall be adjusted on an equitable basis taking into consideration the type and nature of the operations to be licensed and shall be deposited and covered into the Treasury as miscellaneous receipts. There are hereby authorized to be appropriated such funds as Congress may from time to time provide: *Provided*, That there is authorized to be appropriated to the Secretary of Agriculture for enforcement by the Department of Agriculture of the provisions of section 2156 of this title an amount not to exceed $100,000 for the transition quarter ending September 30, 1976, and not to exceed $400,000 for each fiscal year thereafter.

§ 2154 - Effective dates

The regulations referred to in sections 2140 and 2143 of this title shall be prescribed by the Secretary as soon as reasonable but not later than six months from August 24, 1966. Additions and amendments thereto may be prescribed from time to time as may be necessary or advisable. Compliance by dealers with the provisions of this chapter and such regulations shall commence ninety days after the promulgation of such regulations. Compliance by research facilities with the provisions of this chapter and such regulations shall commence six months after the promulgation of such regulations, except that the Secretary may grant extensions of time to research facilities which do not comply with the standards prescribed by the Secretary pursuant to section 2143 of this title provided that the Secretary determines that there is evidence that the research facilities will meet such standards within a reasonable time. Notwithstanding the other provisions of this section, compliance by intermediate handlers, and carriers, and other persons with those provisions of this chapter, as amended by the Animal Welfare Act Amendments of 1976, and those regulations promulgated thereunder, which relate to actions of intermediate handlers and carriers, shall commence 90 days after promulgation of regulations under section 2143 of this title, as amended, with respect to intermediate handlers and carriers, and such regulations shall be promulgated no later than 9 months after April 22, 1976; and compliance by dealers, exhibitors, operators of auction sales, and research facilities with other provisions of this chapter, as so amended, and the regulations thereunder, shall commence upon the expiration of 90 days after April 22, 1976: *Provided*, however, That compliance by all persons with subsections (b), (c), and (d) of section 2143 and with section 2156 of this title, as so amended, shall commence upon the expiration of said ninety-day period. In all other respects, said amendments shall become effective on April 22, 1976.

§ 2155 – Omitted

§ 2156 - Animal fighting venture prohibition
(a) Sponsoring or exhibiting an animal in, attending, or causing an individual who has not attained the age of 16 to attend an animal fighting venture

(1) In general Sponsoring or Exhibiting
Except as provided in paragraph (3), it shall be unlawful for any person to knowingly sponsor or exhibit an animal in an animal fighting venture.

(2) Attending or causing an individual who has not attained the age of 16 to attend – It shall be unlawful for any person to –
knowingly attend an animal fighting venture; or
knowingly cause an individual who has not attained the age of 16 to attend an animal fighting venture.

(3) Special rule for certain States[1]

With respect to fighting ventures involving live birds in a State where it would not be in violation of the law, it shall be unlawful under this subsection for a person to sponsor or exhibit a bird in the fighting venture only if the person knew that any bird in the fighting venture was knowingly bought, sold, delivered, transported, or received in interstate or foreign commerce for the purpose of participation in the fighting venture.

(b) Buying, selling, delivering, possessing, training, or transporting animals for participation in animal fighting venture

It shall be unlawful for any person to knowingly sell, buy, possess, train, transport, deliver, or receive any animal for purposes of having the animal participate in an animal fighting venture.

(c) Use of Postal Service or other interstate instrumentality for promoting or furthering animal fighting venture

It shall be unlawful for any person to knowingly use the mail service of the United States Postal Service or any instrumentality of interstate commerce for commercial speech for purposes of advertising an animal, or an instrument described in subsection (e), for use in an animal fighting venture, promoting[2] or in any other manner furthering an animal fighting venture except as performed outside the limits of the States of the United States.

(d) Violation of State law

Notwithstanding the provisions of subsection (c) of this section, the activities prohibited by such subsection shall be unlawful with respect to fighting ventures involving live birds only if the fight is to take place in a State where it would be in violation of the laws thereof.

(e) Buying, selling, delivering, or transporting sharp instruments for use in animal fighting venture

It shall be unlawful for any person to knowingly sell, buy, transport, or deliver in interstate or foreign commerce a knife, a gaff, or any other sharp instrument attached, or designed or intended to be attached, to the leg of a bird for use in an animal fighting venture.

(f) Investigation of violations by Secretary; assistance by other Federal agencies; issuance of search warrant; forfeiture; costs recoverable in forfeiture or civil action

The Secretary or any other person authorized by him shall make such investigations as the Secretary deems necessary to determine whether any person has violated or is violating any provision of this section, and the Secretary may obtain the assistance of the Federal Bureau of Investigation, the Department of the Treasury, or other law enforcement agencies of the United States, and State and local governmental agencies, in the conduct of such investigations, under cooperative agreements with such agencies. A

---

1 *So in original. Probably should be "States".*
2 *So in original. Probably should be preceded by "or".*

warrant to search for and seize any animal which there is probable cause to believe was involved in any violation of this section may be issued by any judge of the United States or of a State court of record or by a United States magistrate judge within the district wherein the animal sought is located. Any United States marshal or any person authorized under this section to conduct investigations may apply for and execute any such warrant, and any animal seized under such a warrant shall be held by the United States marshal or other authorized person pending disposition thereof by the court in accordance with this subsection. Necessary care including veterinary treatment shall be provided while the animals are so held in custody. Any animal involved in any violation of this section shall be liable to be proceeded against and forfeited to the United States at any time on complaint filed in any United States district court or other court of the United States for any jurisdiction in which the animal is found and upon a judgment of forfeiture shall be disposed of by sale for lawful purposes or by other humane means, as the court may direct. Costs incurred for care of animals seized and forfeited under this section shall be recoverable from the owner of the animals (1) if he appears in such forfeiture proceeding, or (2) in a separate civil action brought in the jurisdiction in which the owner is found, resides, or transacts business.

(g) Definitions

In this section –

(1) the term "**animal fighting venture**" means any event, in or affecting interstate or foreign commerce, that involves a fight conducted or to be conducted between at least 2 animals for purposes of sport, wagering, or entertainment, except that the term "animal fighting venture" shall not be deemed to include any activity the primary purpose of which involves the use of one or more animals in hunting another animal;

(2) the term "**instrumentality of interstate commerce**" means any written, wire, radio, television or other form of communication in, or using a facility of, interstate commerce;

(3) the term "**State**" means any State of the United States, the District of Columbia, the Commonwealth of Puerto Rico, and any territory or possession of the United States;

(4) the term "**animal**" means any live bird, or any live mammal, except man.

(h) Relationship to other provisions

The conduct by any person of any activity prohibited by this section shall not render such person subject to the other sections of this chapter as a dealer, exhibitor, or otherwise.

(i) Conflict with State law

(1) In general

The provisions of this chapter shall not supersede or otherwise invalidate any such State, local, or municipal legislation or ordinance relating to animal fighting ventures except in case of a direct and irreconcilable conflict between any requirements thereunder and this chapter or any rule, regulation, or standard hereunder.

(2) Omitted

(j) Criminal penalties

The criminal penalties for violations of subsection (a), (b), (c), or (e) are provided in section 49 of title 18.

§ 2157 - Release of trade secrets

(a) Release of confidential information prohibited

It shall be unlawful for any member of an Institutional Animal Committee to release any confidential information of the research facility including any information that concerns or relates to –

(1) the trade secrets, processes, operations, style of work, or apparatus; or

(2) the identity, confidential statistical data, amount or source of any income, profits, losses, or expenditures, of the research facility.

(b) Wrongful use of confidential information prohibited

It shall be unlawful for any member of such Committee –

(1) to use or attempt to use to his advantages; or

(2) to reveal to any other person,

any information which is entitled to protection as confidential information under subsection (a) of this section.

(c) Penalties

A violation of subsection (a) or (b) of this section is punishable by –

(1) removal from such Committee; and

(2)(A) a fine of not more than $1,000 and imprisonment of not more than one year; or

(B) if such violation is willful, a fine of not more than $10,000 and imprisonment of not more than three years.

(d) Recovery of damages by injured person; costs; attorney's fee

Any person, including any research facility, injured in its business or property by reason of a violation of this section may recover all actual and consequential damages sustained by such person and the cost of the suit including a reasonable attorney's fee.

(e) Other rights and remedies

Nothing in this section shall be construed to affect any other rights of a person injured in its business or property by reason of a violation of this section. Subsection (d) of this section shall not be construed to limit

AWA

the exercise of any such rights arising out of or relating to a violation of subsections (a) and (b) of this section.

§ 2158 - Protection of pets
(a) Holding period
(1) Requirement
In the case of each dog or cat acquired by an entity described in paragraph (2), such entity shall hold and care for such dog or cat for a period of not less than five days to enable such dog or cat to be recovered by its original owner or adopted by other individuals before such entity sells such dog or cat to a dealer.

(2) Entities described
An entity subject to paragraph (1) is –

    (A) each State, county, or city owned and operated pound or shelter;

    (B) each private entity established for the purpose of caring for animals, such as a humane society, or other organization that is under contract with a State, county, or city that operates as a pound or shelter and that releases animals on a voluntary basis; and

    (C) each research facility licensed by the Department of Agriculture.

(b) Certification
(1) In general
A dealer may not sell, provide, or make available to any individual or entity a random source dog or cat unless such dealer provides the recipient with a valid certification that meets the requirements of paragraph (2) and indicates compliance with subsection (a) of this section.

(2) Requirements
A valid certification shall contain –

    (A) the name, address, and Department of Agriculture license or registration number (if such number exists) of the dealer;

    (B) the name, address, Department of Agriculture license or registration number (if such number exists), and the signature of the recipient of the dog or cat;

    (C) a description of the dog or cat being provided that shall include

        (i) the species and breed or type of such;

        (ii) the sex of such;

        (iii) the date of birth (if known) of such;

        (iv) the color and any distinctive marking of such; and

        (v) any other information that the Secretary by regulation shall determine to be appropriate;

    (D) the name and address of the person, pound, or shelter from which the dog or cat was purchased or otherwise acquired by the dealer, and an assurance that such person, pound, or shelter was notified that such dog or cat may be used for research or educational purposes;

(E) the date of the purchase or acquisition referred to in subparagraph (D);

(F) a statement by the pound or shelter (if the dealer acquired the dog or cat from such) that it satisfied the requirements of subsection (a) of this section; and

(G) any other information that the Secretary of Agriculture by regulation shall determine appropriate.

(3) Records

The original certification required under paragraph (1) shall accompany the shipment of a dog or cat to be sold, provided, or otherwise made available by the dealer, and shall be kept and maintained by the research facility for a period of at least one year for enforcement purposes. The dealer shall retain one copy of the certification provided under this paragraph for a period of at least one year for enforcement purposes.

(4) Transfers

In instances where one research facility transfers animals to another research facility a copy of the certificate must accompany such transfer.

(5) Modification

Certification requirements may be modified to reflect technological advances in identification techniques, such as microchip technology, if the Secretary determines that adequate information such as described in this section, will be collected, transferred, and maintained through such technology.

(c) Enforcement

(1) In general

Dealers who fail to act according to the requirements of this section or who include false information in the certification required under subsection (b) of this section, shall be subject to the penalties provided for under section 2149 of this title.

(2) Subsequent violations

Any dealer who violates this section more than one time shall be subject to a fine of $5,000 per dog or cat acquired or sold in violation of this section.

(3) Permanent revocations

Any dealer who violates this section three or more times shall have such dealers license permanently revoked.

(d) Regulation

Not later than 180 days after November 28, 1990, the Secretary shall promulgate regulations to carry out this section.

§ 2159 - Authority to apply for injunctions

(a) Request

Whenever the Secretary has reason to believe that any dealer, carrier, exhibitor, or intermediate handler is dealing in stolen animals, or is placing the health of any animal in serious danger in violation of this chapter or the regulations or standards promulgated thereunder, the Secretary shall notify the Attorney General, who may apply to the United States district court in which such dealer, carrier, exhibitor, or intermediate handler resides or conducts business for a temporary restraining order or injunction to prevent any such person from operating in violation of this chapter or the regulations and standards prescribed under this chapter.

(b) Issuance

The court shall, upon a proper showing, issue a temporary restraining order or injunction under subsection (a) of this section without bond. Such injunction or order shall remain in effect until a complaint pursuant to section 2149 of this title is issued and dismissed by the Secretary or until an order to cease and desist made thereon by the Secretary has become final and effective or is set aside on appellate review. Attorneys of the Department of Agriculture may, with the approval of the Attorney General, appear in the United States district court representing the Secretary in any action brought under this section.

Animal Welfare Regulations

As of Jan. 1, 2017

As found in the Code of Federal Regulations

Title 9 - Animals and Animal Products
Chapter 1 - Animal and Plant Health Inspection Service,
Department of Agriculture
Subchapter A - Animal Welfare

Parts 1 - 4

ANIMAL WELFARE REGULATIONS

PART 1 – DEFINITION OF TERMS

Authority: 7 U.S.C. 2131-2159; 7 CFR 2.22, 2.80, and 371.7.

§ 1.1 - Definitions.

For the purposes of this subchapter, unless the context otherwise requires, the following terms shall have the meanings assigned to them in this section. The singular form shall also signify the plural and the masculine form shall also signify the feminine. Words undefined in the following paragraphs shall have the meaning attributed to them in general usage as reflected by definitions in a standard dictionary.

AC Regional Director means a veterinarian or his designee, employed by APHIS, who is assigned by the Administrator to supervise and perform the official work of APHIS in a given State or States. As used in part 2 of this subchapter, the AC Regional Director shall be deemed to be the person in charge of the official work of APHIS in the State in which the dealer, exhibitor, research facility, intermediate handler, carrier, or operator of an auction sale has his principal place of business.

Act means the Act of August 24, 1966 (Pub. L. 89-544), (commonly known as the Laboratory Animal Welfare Act), as amended by the Act of December 24, 1970 (Pub. L. 91-579), (the Animal Welfare Act of 1970), the Act of April 22, 1976 (Pub. L. 94-279), (the Animal Welfare Act of 1976), and the Act of December 23, 1985 (Pub. L. 99-198), (the Food Security Act of 1985), and as it may be subsequently amended.

Activity means, for purposes of part 2, subpart C of this subchapter, those elements of research, testing, or teaching procedures that involve the care and use of animals.

Administrative unit means the organizational or management unit at the departmental level of a research facility.

Administrator The Administrator, Animal and Plant Health Inspection Service, or any person authorized to act for the Administrator.

Ambient temperature means the air temperature surrounding the animal.

Animal means any live or dead dog, cat, nonhuman primate, guinea pig, hamster, rabbit, or any other warm-blooded animal, which is being used, or is intended for use for research, teaching, testing, experimentation, or exhibition purposes, or as a pet. This term excludes birds, rats of the genus Rattus, and mice of the genus Mus, bred for use in research; horses not used for research purposes; and other farm animals, such as, but not limited to, livestock or poultry used or intended for use as food or fiber, or livestock or poultry used or intended for use for improving animal nutrition, breeding, management, or production efficiency, or for improving the quality of food or fiber. With

31

respect to a dog, the term means all dogs, including those used for hunting, security, or breeding purposes.

Animal act means any performance of animals where such animals are trained to perform some behavior or action or are part of a show, performance, or exhibition.

APHIS official means any person employed by the Department who is authorized to perform a function under the Act and the regulations in 9 CFR parts 1, 2, and 3.

PART 1
Definitions

Attending veterinarian means a person who has graduated from a veterinary school accredited by the American Veterinary Medical Association's Council on Education, or has a certificate issued by the American Veterinary Medical Association's Education Commission for Foreign Veterinary Graduates, or has received equivalent formal education as determined by the Administrator; has received training and/or experience in the care and management of the species being attended; and who has direct or delegated authority for activities involving animals at a facility subject to the jurisdiction of the Secretary.

Buffer area means that area in a primary enclosure for a swim-with-the-dolphin program that is off-limits to members of the public and that directly abuts the interactive area.

Business hours means a reasonable number of hours between 7 a.m. and 7 p.m., Monday through Friday, except for legal Federal holidays, each week of the year, during which inspections by APHIS may be made.

Business year means the 12-month period during which business is conducted, and may be either on a calendar or fiscal-year basis.

Carrier means the operator of any airline, railroad, motor carrier, shipping line, or other enterprise which is engaged in the business of transporting any animals for hire.

Cat means any live or dead cat (Felis catus) or any cat-hybrid cross.

Class "A" licensee (breeder) means a person subject to the licensing requirements under part 2 and meeting the definition of a "dealer" (§ 1.1), and whose business involving animals consists only of animals that are bred and raised on the premises in a closed or stable colony and those animals acquired for the sole purpose of maintaining or enhancing the breeding colony.

Class "B" licensee means a person subject to the licensing requirements under part 2 and meeting the definition of a "dealer" (§ 1.1), and whose business includes the purchase and/or resale of any animal. This term includes brokers, and operators of an auction sale, as such individuals negotiate or arrange for the purchase, sale, or transport of animals in commerce. Such individuals do not usually take actual physical possession or control of the animals, and do not usually hold animals in any facilities. A class "B" licensee may also exhibit animals as a minor part of the business.

Class "C" licensee (exhibitor) means a person subject to the licensing requirements under part 2 and meeting the definition of an "exhibitor" (§ 1.1), and whose business involves the showing or displaying of animals to the public. A class "C" licensee may buy and sell animals as a minor part of the business in order to maintain or add to his animal collection.

Commerce means trade, traffic, transportation, or other commerce:

(1) Between a place in a State and any place outside of such State, including any foreign country, or between points within the same State but through any place outside thereof, or within any territory, possession, or the District of Columbia; or

(2) Which affects the commerce described in this part.

Committee means the Institutional Animal Care and Use Committee (IACUC) established under section 13(b) of the Act. It shall consist of at least three (3) members, one of whom is the attending veterinarian of the research facility and one of whom is not affiliated in any way with the facility other than as a member of the committee, however, if the research facility has more than one Doctor of Veterinary Medicine (DVM), another DVM with delegated program responsibility may serve. The research facility shall establish the Committee for the purpose of evaluating the care, treatment, housing, and use of animals, and for certifying compliance with the Act by the research facility.

Dealer means any person who, in commerce, for compensation or profit, delivers for transportation, or transports, except as a carrier, buys, or sells, or negotiates the purchase or sale of: Any dog or other animal whether alive or dead (including unborn animals, organs, limbs, blood, serum, or other parts) for research, teaching, testing, experimentation, exhibition, or for use as a pet; or any dog at the wholesale level for hunting, security, or breeding purposes. This term does not include: A retail pet store, as defined in this section; any retail outlet where dogs are sold for hunting, breeding, or security purposes; or any person who does not sell or negotiate the purchase or sale of any wild or exotic animal, dog, or cat and who derives no more than $500 gross income from the sale of animals other than wild or exotic animals, dogs, or cats during any calendar year.

Department means the U.S. Department of Agriculture.

Deputy Administrator means the Deputy Administrator for Animal Care (AC) or any other official of AC to whom authority has been delegated to act in his stead.

Dog means any live or dead dog (Canis familiaris) or any dog-hybrid cross.

PART 1

Definitions

Dwarf hamster means any species of hamster such as the Chinese and Armenian species whose adult body size is substantially less than that attained by the Syrian or Golden species of hamsters.

Endangered species means those species defined in the Endangered Species Act (16 U.S.C. 1531 et seq.) and as it may be subsequently amended.

Euthanasia means the humane destruction of an animal accomplished by a method that produces rapid unconsciousness and subsequent death without evidence of pain or distress, or a method that utilizes anesthesia produced by an agent that causes painless loss of consciousness and subsequent death.

PART 1
Definitions

Exhibitor means any person (public or private) exhibiting any animals, which were purchased in commerce or the intended distribution of which affects commerce, or will affect commerce, to the public for compensation, as determined by the Secretary. This term includes carnivals, circuses, animal acts, zoos, and educational exhibits, exhibiting such animals whether operated for profit or not. This term excludes retail pet stores, horse and dog races, organizations sponsoring and all persons participating in State and county fairs, livestock shows, rodeos, field trials, coursing events, purebred dog and cat shows and any other fairs or exhibitions intended to advance agricultural arts and sciences as may be determined by the Secretary.

Exotic animal means any animal not identified in the definition of "animal" provided in this part that is native to a foreign country or of foreign origin or character, is not native to the United States, or was introduced from abroad. This term specifically includes animals such as, but not limited to, lions, tigers, leopards, elephants, camels, antelope, anteaters, kangaroos, and water buffalo, and species of foreign domestic cattle, such as Ankole, Gayal, and Yak.

Farm animal means any domestic species of cattle, sheep, swine, goats, llamas, or horses, which are normally and have historically, been kept and raised on farms in the United States, and used or intended for use as food or fiber, or for improving animal nutrition, breeding, management, or production efficiency, or for improving the quality of food or fiber. This term also includes animals such as rabbits, mink, and chinchilla, when they are used solely for purposes of meat or fur, and animals such as horses and llamas when used solely as work and pack animals.

Federal agency means an Executive agency as such term is defined in section 105 of title 5, United States Code, and with respect to any research facility means the agency from which the research facility receives a Federal award for the conduct of research, experimentation, or testing involving the use of animals.

Federal award means any mechanism (including a grant, award, loan, contract, or cooperative agreement) under which Federal funds are used to support the conduct of research, experimentation, or testing, involving the use of animals. The permit system established under the authorities of the Endangered Species Act, the Marine Mammal Protection Act, and the Migratory Bird Treaty Act, are not considered to be Federal awards under the Animal Welfare Act.

Federal research facility means each department, agency, or instrumentality of the United States which uses live animals for research or experimentation.

Field study means a study conducted on free-living wild animals in their natural habitat. However, this term excludes any study that involves an invasive procedure, harms, or materially alters the behavior of an animal under study.

Handling means petting, feeding, watering, cleaning, manipulating, loading, crating, shifting, transferring, immobilizing, restraining, treating, training, working and moving, or any similar activity with respect to any animal.

Housing facility means any land, premises, shed, barn, building, trailer, or other structure or area housing or intended to house animals.

Hybrid cross means an animal resulting from the crossbreeding between two different species or types of animals. Crosses between wild animal species, such as lions and tigers, are considered to be wild animals. Crosses between wild animal species and domestic animals, such as dogs and wolves or buffalo and domestic cattle, are considered to be domestic animals.

Impervious surface means a surface that does not permit the absorption of fluids. Such surfaces are those that can be thoroughly and repeatedly cleaned and disinfected, will not retain odors, and from which fluids bead up and run off or can be removed without their being absorbed into the surface material.

Indoor housing facility means any structure or building with environmental controls housing or intended to house animals and meeting the following three requirements:

(1) It must be capable of controlling the temperature within the building or structure within the limits set forth for that species of animal, of maintaining humidity levels of 30 to 70 percent and of rapidly eliminating odors from within the building; and

(2) It must be an enclosure created by the continuous connection of a roof, floor, and walls (a shed or barn set on top of the ground does not have a continuous connection between the walls and the ground unless a foundation and floor are provided); and

(3) It must have at least one door for entry and exit that can be opened and closed (any windows or openings which provide natural light must be covered with a transparent material such as glass or hard plastic).

Interactive area means that area in a primary enclosure for a swim-with-the-dolphin program where an interactive session takes place.

Interactive session means a swim-with-the-dolphin program session where members of the public enter a primary enclosure to interact with cetaceans.

Intermediate handler means any person, including a department, agency, or instrumentality of the United States or of any State or local government (other than a dealer, research facility, exhibitor, any person excluded from the definition of a dealer, research facility, or exhibitor, an operator of an auction sale, or a carrier), who is engaged in any business in which he receives custody of animals in connection with their transportation in commerce.

Inspector means any person employed by the Department who is authorized to perform a function under the Act and the regulations in 9 CFR parts 1, 2, and 3.

Institutional official means the individual at a research facility who is authorized to legally commit on behalf of the research facility that the requirements of 9 CFR parts 1, 2, and 3 will be met.

Isolation in regard to marine mammals means the physical separation of animals to prevent contact and a separate, noncommon, water circulation and filtration system for the isolated animals.

Licensed veterinarian means a person who has graduated from an accredited school of veterinary medicine or has received equivalent formal education as determined by the Administrator, and who has a valid license to practice veterinary medicine in some State.

Licensee means any person licensed according to the provisions of the Act and the regulations in part 2 of this subchapter.

Major operative procedure means any surgical intervention that penetrates and exposes a body cavity or any procedure which produces permanent impairment of physical or physiological functions.

Minimum horizontal dimension (MHD) means the diameter of a circular pool of water, or in the case of a square, rectangle, oblong, or other shape pool, the diameter of the largest circle that can be inserted within the confines of such a pool of water.

Mobile or traveling housing facility means a transporting vehicle such as a truck, trailer, or railway car, used to house animals while traveling for exhibition or public education purposes.

Non-conditioned animals means animals which have not been subjected to special care and treatment for sufficient time to stabilize, and where necessary, to improve their health.

Nonhuman primate means any nonhuman member of the highest order of mammals including prosimians, monkeys, and apes.

Operator of an auction sale means any person who is engaged in operating an auction at which animals are purchased or sold in commerce.

Outdoor housing facility means any structure, building, land, or premise, housing or intended to house animals, which does not meet the definition of any other type of housing facility provided in the regulations, and in which temperatures cannot be controlled within set limits.

Painful procedure as applied to any animal means any procedure that would reasonably be expected to cause more than slight or momentary pain or distress in a human being to which that procedure was applied, that is, pain in excess of that caused by injections or other minor procedures.

Paralytic drug means a drug which causes partial or complete loss of muscle contraction and which has no anesthetic or analgesic properties, so that the animal cannot move, but is completely aware of its surroundings and can feel pain.

PART 1
Definitions

Person means any individual, partnership, firm, joint stock company, corporation, association, trust, estate, or other legal entity.

Pet animal means any animal that has commonly been kept as a pet in family households in the United States, such as dogs, cats, guinea pigs, rabbits, and hamsters. This term excludes exotic animals and wild animals.

Positive physical contact means petting, stroking, or other touching, which is beneficial to the well-being of the animal.

Pound or shelter means a facility that accepts and/or seizes animals for the purpose of caring for them, placing them through adoption, or carrying out law enforcement, whether or not the facility is operated for profit.

Primary conveyance means the main method of transportation used to convey an animal from origin to destination, such as a motor vehicle, plane, ship, or train.

Primary enclosure means any structure or device used to restrict an animal or animals to a limited amount of space, such as a room, pen, run, cage, compartment, pool, or hutch.

Principal investigator means an employee of a research facility, or other person associated with a research facility, responsible for a proposal to conduct research and for the design and implementation of research involving animals.

Quorum means a majority of the Committee members.

Random source means dogs and cats obtained from animal pounds or shelters, auction sales, or from any person who did not breed and raise them on his or her premises.

Registrant means any research facility, carrier, intermediate handler, or exhibitor not required to be licensed under section 3 of the Act, registered pursuant to the provisions of the Act and the regulations in part 2 of this subchapter.

Research facility means any school (except an elementary or secondary school), institution, organization, or person that uses or intends to use live animals in research, tests, or experiments, and that (1) purchases or transports live animals in commerce, or (2) receives funds under a grant, award, loan, or contract from a department, agency, or instrumentality of the United States for the purpose of carrying out research, tests, or experiments: *Provided,* That the Administrator may exempt, by regulation, any such school, institution, organization, or person that does not use or intend to use live dogs or cats, except those schools, institutions, organizations, or persons, which use substantial numbers (as determined by the Administrator) of live animals the principal function of which schools, institutions, organizations, or persons, is biomedical research or testing, when in the judgment of the Administrator, any such exemption does not vitiate the purpose of the Act.

PART 1
Definitions

Retail pet store means a place of business or residence at which the seller, buyer, and the animal available for sale are physically present so that every buyer may personally observe the animal prior to purchasing and/or taking custody of that animal after purchase, and where only the following animals are sold or offered for sale, at retail, for use as pets: Dogs, cats, rabbits, guinea pigs, hamsters, gerbils, rats, mice, gophers, chinchilla, domestic ferrets, domestic farm animals, birds, and coldblooded species. In addition to persons that meet these criteria, *retail pet store* also includes any person who meets the criteria in § 2.1(a)(3)(vii) of this subchapter. Such definition excludes –

(1) Establishments or persons who deal in dogs used for hunting, security, or breeding purposes;

(2) Establishments or persons, except those that meet the criteria in § 2.1(a)(3)(vii), exhibiting, selling, or offering to exhibit or sell any wild or exotic or other non-pet species of warm-blooded animals (except birds), such as skunks, raccoons, nonhuman primates, squirrels, ocelots, foxes, coyotes, etc.;

(3) Any establishment or person selling warm-blooded animals (except birds, and laboratory rats and mice) for research or exhibition purposes;

(4) Any establishment wholesaling any animals (except birds, rats and mice); and

(5) Any establishment exhibiting pet animals in a room that is separate from or adjacent to the retail pet store, or in an outside area, or anywhere off the retail pet store premises.

Sanctuary area means that area in a primary enclosure for a swim-with-the-dolphin program that is off-limits to the public and that directly abuts the buffer area.

Sanitize means to make physically clean and to remove and destroy, to the maximum degree that is practical, agents injurious to health.

Secretary means the Secretary of Agriculture of the United States or his representative who shall be an employee of the Department.

Sheltered housing facility means a housing facility which provides the animals with shelter; protection from the elements; and protection from temperature extremes at all times. A sheltered housing facility may consist of runs or pens totally enclosed in a barn or building, or of connecting inside/outside runs or pens with the inside pens in a totally enclosed building.

Standards means the requirements with respect to the humane housing, exhibition, handling, care, treatment, temperature, and transportation of animals by dealers, exhibitors research facilities, carriers, intermediate handlers, and operators of auction sales as set forth in part 3 of this subchapter.

PART 1
Definitions

State means a State of the United States, the District of Columbia, Commonwealth of Puerto Rico, the Virgin Islands, Guam, American Samoa, or any other territory or possession of the United States.

Study area means any building room, area, enclosure, or other containment outside of a core facility or centrally designated or managed area in which animals are housed for more than 12 hours.

Swim-with-the-dolphin (SWTD) program means any human-cetacean interactive program in which a member of the public enters the primary enclosure in which an SWTD designated cetacean is housed to interact with the animal. This interaction includes, but such inclusions are not limited to, wading, swimming, snorkeling, or scuba diving in the enclosure. This interaction excludes, but such exclusions are not limited to, feeding and petting pools, and the participation of any member(s) of the public audience as a minor segment of an educational presentation or performance of a show.

Transporting device means an interim vehicle or device, other than man, used to transport an animal between the primary conveyance and the terminal facility or in and around the terminal facility of a carrier or intermediate handler.

Transporting vehicle means any truck, car, trailer, airplane, ship, or railroad car used for transporting animals.

Weaned means that an animal has become accustomed to take solid food and has so done, without nursing, for a period of at least 5 days.

Wild animal means any animal which is now or historically has been found in the wild, or in the wild state, within the boundaries of the United States, its territories, or possessions. This term includes, but is not limited to, animals such as: deer, skunk, opossum, raccoon, mink, armadillo, coyote, squirrel, fox, wolf, etc.

Wild state means living in its original, natural condition; not domesticated.

Zoo means any park, building, cage, enclosure, or other structure or premise in which a live animal or animals are kept for public exhibition or viewing, regardless of compensation.

[54 FR 36119, Aug. 31, 1989, as amended at 55 FR 12631, Apr. 5, 1990; 58 FR 39129, July 22, 1993; 62 FR 43275, Aug. 13, 1997; 63 FR 47148, Sept. 4, 1998; 63 FR 62926, Nov. 10, 1998; 65 FR 6314, Feb. 9, 2000; 68 FR 12285, Mar. 14, 2003; 69 FR 31514, June 4, 2004; 69 FR 42099, July 14, 2004; 78 FR 57227, Sept. 18, 2013]

PART 1
Definitions

Effective Date Note: At 64 FR 15920, Apr. 2, 1999, the definitions of buffer area, interactive area, interactive session, sanctuary area, and swim-with-the-dolphin (SWTD) program were suspended, effective Apr. 2, 1999.

PART 2 – REGULATIONS

Subpart A – Licensing

Subpart B – Registration

Subpart C – Research Facilities

Subpart D – Attending Veterinarian and Adequate Veterinary Care

Subpart E – Identification of Animals

PART 2
Table of Contents

Authority: 7 U.S.C. 2131-2159; 7 CFR 2.22, 2.80, and 371.7.

Source: 54 FR 36147, Aug. 31, 1989, unless otherwise noted.

PART 2
Table of Contents

Subpart A – Licensing

§ 2.1 - Requirements and application.

(a)(1) Any person operating or intending to operate as a dealer, exhibitor, or operator of an auction sale, except persons who are exempted from the licensing requirements under paragraph (a)(3) of this section, must have a valid license. A person must be 18 years of age or older to obtain a license. A person seeking a license shall apply on a form which will be furnished by the AC Regional Director in the State in which that person operates or intends to operate. The applicant shall provide the information requested on the application form, including a valid mailing address through which the licensee or applicant can be reached at all times, and a valid premises address where animals, animal facilities, equipment, and records may be inspected for compliance. The applicant shall file the completed application form with the AC Regional Director.

(2) If an applicant for a license or license renewal operates in more than one State, he or she shall apply in the State in which he or she has his or her principal place of business. All premises, facilities, or sites where such person operates or keeps animals shall be indicated on the application form or on a separate sheet attached to it. The completed application form, along with the application fee indicated in paragraph (c) of this section, and the annual license fee indicated in table 1 or 2 of § 2.6 shall be filed with the AC Regional Director.

PART 2
Subpart A

(3) The following persons are exempt from the licensing requirements under section 2 or section 3 of the Act:

(i) Retail pet stores as defined in part 1 of this subchapter;

(ii) Any person who sells or negotiates the sale or purchase of any animal except wild or exotic animals, dogs, or cats, and who derives no more than $500 gross income from the sale of such animals during any calendar year and is not otherwise required to obtain a license;

(iii) Any person who maintains a total of four or fewer breeding female dogs, cats, and/or small exotic or wild mammals, such as hedgehogs, degus, spiny mice, prairie dogs, flying squirrels, and jerboas, and who sells, at wholesale, only the offspring of these dogs, cats, and/or small exotic or wild mammals, which were born and raised on his or her premises, for pets or exhibition, and is not otherwise required to obtain a license. This exemption does not extend to any person residing in a household that collectively maintains a total of more than four breeding female dogs, cats, and/or small exotic or wild mammals, regardless of ownership, nor to any person maintaining breeding female dogs, cats, and/or small exotic or wild mammals on premises on which more than four breeding female dogs, cats, and/or small exotic or wild mammals are maintained, nor to any person acting in concert with others where they collectively maintain a total of more than four breeding female dogs, cats, and/or small exotic or wild mammals regardless of ownership;

(iv) Any person who sells fewer than 25 dogs and/or cats per year, which were born and raised on his or her premises, for research, teaching, or testing purposes or to any research facility and is not otherwise required to obtain a license. This exemption does not extend to any person residing in a household that collectively sells 25 or more dogs and/or cats, regardless of ownership, nor to any person acting in concert with others where they collectively sell 25 or more dogs and/or cats, regardless of ownership. The sale of any dog or cat not born and raised on the premises for research purposes requires a license;

(v) Any person who arranges for transportation or transports animals solely for the purpose of breeding, exhibiting in purebred shows, boarding (not in association with commercial transportation), grooming, or medical treatment, and is not otherwise required to obtain a license;

(vi) Any person who buys, sells, transports, or negotiates the sale, purchase, or transportation of any animals used only for the purposes of food or fiber (including fur);

(vii) Any person including, but not limited to, purebred dog or cat fanciers, who maintains a total of four or fewer breeding female dogs, cats, and/or small exotic or wild mammals, such as hedgehogs, degus, spiny mice, prairie dogs, flying squirrels, and jerboas, and who sells, at retail, only the offspring of these dogs, cats, and/or small exotic or wild mammals, which were born and raised on his or her premises, for pets or exhibition, and is not otherwise required to obtain a license. This exemption does not extend to any person residing in a household that collectively maintains a total of more than four breeding female dogs, cats, and/or small exotic or wild mammals, regardless of ownership, nor to any person maintaining breeding female dogs, cats, and/or small exotic or wild mammals on premises on which more than four breeding female dogs, cats, and/or small exotic or wild mammals are maintained, nor to any person acting in concert with others where they collectively maintain a total of more than four breeding female dogs, cats, and/or small exotic or wild mammals regardless of ownership;

(viii) Any person who buys animals solely for his or her own use or enjoyment and does not sell or exhibit animals, or is not otherwise required to obtain a license;

(b) No person shall have more than one license.

(c) A license will be issued to any applicant, except as provided in §§ 2.10 and 2.11, when:

(1) The applicant has met the requirements of this section and §§ 2.2 and 2.3; and

(2) The applicant has paid the application fee of $10 and the annual license fee indicated in § 2.6 to the appropriate Animal Care regional office for an initial license, and, in the case of a license renewal, the annual license fee has been received by the appropriate Animal Care regional office on or before the expiration date of the license.

(d)(1) A licensee who wishes a renewal must submit to the appropriate Animal Care regional office a completed application form and the annual license fee indicated in § 2.6 by certified check, cashier's check, personal check, money order, or credit card. The application form and the annual license fee must be received by the appropriate Animal Care regional office on or before the expiration date of the license. An applicant whose check is returned by the bank will be charged a fee of $20 for each returned check. A returned check will be deemed nonpayment of fee and will result in the denial of the license. If an applicant's check is returned, subsequent fees must be paid by certified check, cashier's check, or money order.

(2) A license fee indicated in § 2.6 must also be paid if an applicant is applying for a changed class of license. The applicant may pay the fee by certified check, cashier's check, personal check, money order, or credit card. An applicant whose check is returned by a bank will be charged a fee of $20 for each returned check. If an applicant's check is returned, subsequent fees must be paid by certified check, cashier's check, or money order.

PART 2
Subpart A

(e) The failure of any person to comply with any provision of the Act, or any of the provisions of the regulations or standards in this subchapter, shall constitute grounds for denial of a license; or for its suspension or revocation by the Secretary, as provided in the Act.

(Approved by the Office of Management and Budget under control numbers 0579-0254 and 0579-0392)

[54 FR 36147, Aug. 31, 1989, as amended at 63 FR 62926, Nov. 10, 1998; 69 FR 42099, July 14, 2004; 78 FR 57227, Sept. 18, 2013]

§ 2.2 - Acknowledgement of regulations and standards.

(a) *Application for initial license.* APHIS will supply a copy of the applicable regulations and standards to the applicant with each request for a license application. The applicant shall acknowledge receipt of the regulations and standards and agree to comply with them by signing the application form before a license will be issued.

(b) *Application for license renewal.* APHIS will renew a license after the applicant certifies by signing the application form that, to the best of the applicant's knowledge and belief, he or she is in compliance with the regulations and standards and agrees to continue to comply with the regulations and standards. APHIS will supply a copy of the applicable regulations and standards to the applicant upon request.

(Approved by the Office of Management and Budget under control number 0579-0254)

[60 FR 13895, Mar. 15, 1995, as amended at 69 FR 42100, July 14, 2004]

§ 2.3 - Demonstration of compliance with standards and regulations.

(a) Each applicant must demonstrate that his or her premises and any animals, facilities, vehicles, equipment, or other premises used or intended for use in the business comply with the regulations and standards set forth in parts 2 and 3 of this subchapter. Each applicant for an initial license or license renewal must make his or her animals, premises, facilities, vehicles, equipment, other premises, and records available for inspection during business hours and at other times mutually agreeable to the applicant and APHIS, to ascertain the applicant's compliance with the standards and regulations.

PART 2
Subpart A

(b) Each applicant for an initial license must be inspected by APHIS and demonstrate compliance with the regulations and standards, as required in paragraph (a) of this section, before APHIS will issue a license. If the first inspection reveals that the applicant's animals, premises, facilities, vehicles, equipment, other premises, or records do not meet the requirements of this subchapter, APHIS will advise the applicant of existing deficiencies and the corrective measures that must be completed to come into compliance with the regulations and standards. An applicant who fails the first inspection will have two additional chances to demonstrate his or her compliance with the regulations and standards through a second inspection by APHIS. The applicant must request the second inspection, and if applicable, the third inspection, within 90 days following the first inspection. If the applicant fails inspection or fails to request re-inspections within the 90-day period, he or she will forfeit the application fee and cannot reapply for a license for a period of 6 months from the date of the failed third inspection or the expiration of the time to request a third inspection. Issuance of a license will be denied until the applicant demonstrates upon inspection that the animals, premises, facilities, vehicles, equipment, other premises, and records are in compliance with all regulations and standards in this subchapter.

(Approved by the Office of Management and Budget under control number 0579-0254)

[54 FR 36147, Aug. 31, 1989, as amended at 69 FR 42100, July 14, 2004]

§ 2.4 - Non-interference with APHIS officials.

A licensee or applicant for an initial license shall not interfere with, threaten, abuse (including verbally abuse), or harass any APHIS official in the course of carrying out his or her duties.

§ 2.5 - Duration of license and termination of license.

(a) A license issued under this part shall be valid and effective unless:

(1) The license has been revoked or suspended pursuant to section 19 of the Act.

(2) The license is voluntarily terminated upon request of the licensee, in writing, to the AC Regional Director.

(3) The license has expired or been terminated under this part.

(4) The annual license fee has not been paid to the appropriate Animal Care regional office as required. There will not be a refund of the annual license fee if a license is terminated prior to its expiration date.

(b) Any person who is licensed must file an application for a license renewal and an annual report form (APHIS Form 7003), as required by § 2.7 of this part, and pay the required annual license fee. The required annual license fee must be received in the appropriate Animal Care regional office on or before the expiration date of the license or the license will expire and automatically terminate. Failure to comply with the annual reporting requirements or pay the required annual license fee on or before the expiration date of the license will result in automatic termination of the license.

PART 2
Subpart A

(c) Any person who seeks the reinstatement of a license that has been automatically terminated must follow the procedure applicable to new applicants for a license set forth in § 2.1.

(d) Licenses are issued to specific persons for specific premises and do not transfer upon change of ownership, nor are they valid at a different location.

(e) A license which is invalid under this part shall be surrendered to the AC Regional Director. If the license cannot be found, the licensee shall provide a written statement so stating to the AC Regional Director.

[54 FR 36147, Aug. 31, 1989, as amended at 60 FR 13895, Mar. 15, 1995; 63 FR 62926, Nov. 10, 1998; 69 FR 42100, July 14, 2004]

§ 2.6 - Annual license fees.

(a) For an initial license, the applicant must submit a $10 application fee in addition to the initial license fee prescribed in this section. Licensees applying for license renewal or changed class of license must submit only the license fee prescribed in this section. The license fee for an initial license,

license renewal, or changed class of license is determined from table 1 or 2 in paragraph (c) of this section. Paragraph (b) of this section indicates the method used to calculate the license fee. All initial license and changed class of license fees must be submitted to the appropriate Animal Care regional office, and, in the case of license renewals, all fees must be received by the appropriate Animal Care regional office on or before the expiration date of the license.

(b)(1) Class "A" license. The annual license renewal fee for a Class "A" dealer shall be based on 50 percent of the total gross amount, expressed in dollars, derived from the sale of animals to research facilities, dealers, exhibitors, retail pet stores, and persons for use as pets, directly or through an auction sale, by the dealer or applicant during his or her preceding business year (calendar or fiscal) in the case of a person who operated during such a year. If animals are leased, the lessor shall pay a fee based on 50 percent of any compensation received from the leased animals and the lessee shall pay a fee based upon the net compensation received from the leased animals, as indicated for dealers in Table 1 in paragraph (c) of this section.

PART 2
Subpart A

(2) Class "B" license. The annual license renewal fee for a Class "B" dealer shall be established by calculating the total amount received from the sale of animals to research facilities, dealers, exhibitors, retail pet stores, and persons for use as pets, directly or through an auction sale, during the preceding business year (calendar or fiscal) less the amount paid for the animals by the dealer or applicant. This net difference, exclusive of other costs, shall be the figure used to determine the license fee of a Class "B" dealer.

If animals are leased, the lessor and lessee shall each pay a fee based on the net compensation received from the leased animals calculated from Table 1 in paragraph (c) of this section.

(3) The annual license renewal fee for a broker or operator of an auction sale shall be that of a class "B" dealer and shall be based on the total gross amount, expressed in dollars, derived from commissions or fees charged for the sale of animals, or for negotiating the sale of animals, by brokers or by the operator of an auction sale, to research facilities, dealers, exhibitors, retail pet stores, and persons for use as pets, during the preceding business year (calendar or fiscal).

(4) In the case of a new applicant for a license as a dealer, broker or operator of an auction sale who did not operate during a preceding business year, the annual license fee will be based on the anticipated yearly dollar amount of business, as provided in paragraphs (b)(1), (2), and (3) of this section, derived from the sale of animals to research facilities, dealers, exhibitors, retail pet stores, and persons for use as pets, directly or through an auction sale.

(5) The amount of the annual fee to be paid upon application for a class "C" license as an exhibitor under this section shall be based on the number of animals which the exhibitor owned, held, or exhibited at the time the application is signed and dated or during the previous year, whichever is greater, and will be the amount listed in Table 2 in paragraph (c) of this section. Animals which are leased shall be included in the number of animals being held by both the lessor and the lessee when calculating the annual fee. An exhibitor shall pay his or her annual license fee on or before the expiration date of the license and the fee shall be based on the number of animals which the exhibitor is holding or has held during the year (both owned and leased).

(c) The license fee shall be computed in accordance with the following tables:

TABLE 1 – DEALERS, BROKERS, AND OPERATORS OF AN AUCTION SALE – CLASS "A" and "B" LICENSE

Over	But not over	Initial license fee	Annual or changed class of license fee
$0	$500	$30	$40
500	2,000	60	70
2,000	10,000	120	130
10,000	25,000	225	235
25,000	50,000	350	360
50,000	100,000	475	485
100,000		750	760

TABLE 2 – EXHIBITORS – CLASS "C" LICENSE

Number of animals	Initial license fee	Annual or changed class of license
1 to 5	$30	$40
6 to 25	75	85
26 to 50	175	185
51 to 500	225	235
501 and up	300	310

(d) If a person meets the licensing requirements for more than one class of license, he shall be required to obtain a license and pay the fee for the type business which is predominant for his operation, as determined by the Secretary.

(e) In any situation in which a licensee shall have demonstrated in writing to the satisfaction of the Secretary that he or she has good reason to believe that the dollar amount of his or her business for the forthcoming business year will be less than the previous business year, then his or her estimated dollar amount of business shall be used for computing the license fee for the forthcoming business year: *Provided, however,* That if the dollar amount upon which the license fee is based for that year does in fact exceed the amount estimated, the difference in amount of the fee paid and that which was due under paragraphs (b) and (c) of this section based upon the actual dollar business upon which the license fee is based, shall be payable in addition to the required annual license fee for the next subsequent year, on the anniversary date of his or her license as prescribed in this section.

[54 FR 36147, Aug. 31, 1989, as amended at 63 FR 62926, Nov. 10, 1998; 69 FR 42101, July 14, 2004]

§ 2.7 - Annual report by licensees.

(a) Each year, within 30 days prior to the expiration date of his or her license, a licensee shall file with the AC Regional Director an application for license renewal and annual report upon a form which the AC Regional Director will furnish to him or her upon request.

(b) A person licensed as a dealer shall set forth in his or her license renewal application and annual report the dollar amount of business, from the sale of animals, upon which the license fee is based, directly or through an auction sale, to research facilities, dealers, exhibitors, retail pet stores, and persons for use as pets, by the licensee during the preceding business year (calendar or fiscal), and any other information as may be required thereon.

(c) A licensed dealer who operates as a broker or an operator of an auction sale shall set forth in his or her license renewal application and annual report the total gross amount, expressed in dollars, derived from commissions or fees charged for the sale of animals by the licensee to research facilities, dealers, exhibitors, retail pet stores, and persons for use as pets, during the preceding business year (calendar or fiscal), and any other information as may be required thereon.

(d) A person licensed as an exhibitor shall set forth in his or her license renewal application and annual report the number of animals owned, held, or exhibited by him or her, including those which are leased, during the previous year or at the time he signs and dates the report, whichever is greater.

[54 FR 36147, Aug. 31, 1989, as amended at 63 FR 62926, Nov. 10, 1998]

§ 2.8 - Notification of change of name, address, control, or ownership of business.

A licensee shall promptly notify the AC Regional Director by certified mail of any change in the name, address, management, or substantial control or ownership of his business or operation, or of any additional sites, within 10 days of any change.

[54 FR 36147, Aug. 31, 1989, as amended at 63 FR 62926, Nov. 10, 1998]

§ 2.9 - Officers, agents, and employees of licensees whose licenses have been suspended or revoked.

Any person who has been or is an officer, agent, or employee of a licensee whose license has been suspended or revoked and who was responsible for or participated in the violation upon which the order of suspension or revocation was based will not be licensed within the period during which the order of suspension or revocation is in effect.

§ 2.10 - Licensees whose licenses have been suspended or revoked.

(a) Any person whose license has been suspended for any reason shall not be licensed in his or her own name or in any other manner within the period during which the order of suspension is in effect. No partnership, firm, corporation, or other legal entity in which any such person has a substantial interest, financial or otherwise, will be licensed during that period. Any person whose license has been suspended for any reason may apply to the AC Regional Director, in writing, for reinstatement of his or her license. No license will be renewed during the period that it is suspended. Renewal of the license may be initiated during the suspension in accordance with §§ 2.2(b) and 2.12.

(b) Any person whose license has been revoked shall not be licensed in his or her own name or in any other manner; nor will any partnership, firm, corporation, or other legal entity in which any such person has a substantial interest, financial or otherwise, be licensed.

(c) Any person whose license has been suspended or revoked shall not buy, sell, transport, exhibit, or deliver for transportation, any animal during the period of suspension or revocation.

[54 FR 36147, Aug. 31, 1989, as amended at 63 FR 62926, Nov. 10, 1998; 69 FR 42101, July 14, 2004]

§ 2.11 - Denial of initial license application.

(a) A license will not be issued to any applicant who:

(1) Has not complied with the requirements of §§ 2.1, 2.2, 2.3, and 2.4 and has not paid the fees indicated in § 2.6;

(2) Is not in compliance with any of the regulations or standards in this subchapter;

(3) Has had a license revoked or whose license is suspended, as set forth in § 2.10;

(4) Has pled *nolo contendere* (no contest) or has been found to have violated any Federal, State, or local laws or regulations pertaining to animal cruelty within 1 year of application, or after 1 year if the Administrator determines that the circumstances render the applicant unfit to be licensed;

(5) Is or would be operating in violation or circumvention of any Federal, State, or local laws; or

(6) Has made any false or fraudulent statements or provided any false or fraudulent records to the Department or other government agencies, or has pled *nolo contendere* (no contest) or has been found to have violated any Federal, State, or local laws or regulations pertaining to the transportation, ownership, neglect, or welfare of animals, or is otherwise unfit to be licensed and the Administrator determines that the issuance of a license would be contrary to the purposes of the Act.

PART 2
Subpart A

(b) An applicant whose license application has been denied may request a hearing in accordance with the applicable rules of practice for the purpose of showing why the application for license should not be denied. The license denial shall remain in effect until the final legal decision has been rendered. Should the license denial be upheld, the applicant may again apply for a license 1 year from the date of the final order denying the application, unless the order provides otherwise.

(c) No partnership, firm, corporation, or other legal entity in which a person whose license application has been denied has a substantial interest, financial or otherwise, will be licensed within 1 year of the license denial.

(d) No license will be issued under circumstances that the Administrator determines would circumvent any order suspending, revoking, terminating, or denying a license under the Act.

[54 FR 36147, Aug. 31, 1989, as amended at 69 FR 42101, July 14, 2004]

§ 2.12 - Termination of a license.

A license may be terminated during the license renewal process or at any other time for any reason that an initial license application may be denied pursuant to § 2.11 after a hearing in accordance with the applicable rules of practice.

[69 FR 42101, July 14, 2004]

Subpart B – Registration

§ 2.25 - Requirements and procedures.

(a) Each carrier and intermediate handler, and each exhibitor not required to be licensed under section 3 of the Act and the regulations of this subchapter, shall register with the Secretary by completing and filing a properly executed form which will be furnished, upon request, by the AC Regional Director. The registration form shall be filed with the AC Regional Director for the State in which the registrant has his or her principal place of business, and shall be updated every 3 years by the completion and filing of a new registration form which will be provided by the AC Regional Director

(b) A subsidiary of a business corporation, rather than the parent corporation, will be registered as an exhibitor unless the subsidiary is under such direct control of the parent corporation that the Secretary determines that it is necessary that the parent corporation be registered to effectuate the purposes of the Act.

(c) No registrant or person required to be registered shall interfere with, threaten, abuse (including verbally abuse), or harass any APHIS official who is in the course of carrying out his or her duties.

[54 FR 36147, Aug. 31, 1989, as amended at 63 FR 62926, Nov. 10, 1998; 69 FR 42101, July 14, 2004]

PART 2
Subpart B

§ 2.26 - Acknowledgment of regulations and standards.

APHIS will supply a copy of the regulations and standards in this subchapter with each registration form. The registrant shall acknowledge receipt of and shall agree to comply with the regulations and standards by signing a form provided for this purpose by APHIS, and by filing it with the AC Regional Director.

[54 FR 36147, Aug. 31, 1989, as amended at 63 FR 62926, Nov. 10, 1998]

§ 2.27 - Notification of change of operation.

(a) A registrant shall notify the AC Regional Director by certified mail of any change in the name, address, or ownership, or other change in operations affecting its status as an exhibitor, carrier, or intermediate handler, within 10 days after making such change.

(b)(1) A registrant which has not used, handled, or transported animals for a period of at least 2 years may be placed in an inactive status by making a written request to the AC Regional Director a registrant shall notify the AC Regional Director in writing at least 10 days before using, handling, or transporting animals again after being in an inactive status.

(2) A registrant which goes out of business or which ceases to function as a carrier, intermediate handler, or exhibitor, or which changes its method of operation so that it no longer uses, handles, or transports animals, and which does not plan to use, handle, or transport animals again at any time in the future, may have its registration canceled by making a written request to the AC Regional Director. The former registrant is responsible for reregistering and demonstrating its compliance with the Act and regulations should it start using, handling, or transporting animals at any time after its registration is canceled.

[54 FR 36147, Aug. 31, 1989, as amended at 63 FR 62926, Nov. 10, 1998]

PART 2
Subpart B

Subpart C – Research Facilities

§ 2.30 - Registration.

(a) *Requirements and procedures.*

(1) Each research facility other than a Federal research facility, shall register with the Secretary by completing and filing a properly executed form which will be furnished, upon request, by the AC Regional Director. The registration form shall be filed with the AC Regional Director for the State in which the research facility has its principal place of business, and shall be updated every 3 years by the completion and filing of a new registration form which will be provided by the AC Regional Director. Except as provided in paragraph (a)(2) of this section, where a school or department of a university or college uses or intends to use live animals for research, tests, experiments, or teaching, the university or college rather than the school or department will be considered the research facility and will be required to register with the Secretary. An official who has the legal authority to bind the parent organization shall sign the registration form.

(2) In any situation in which a school or department of a university or college demonstrates to the Secretary that it is a separate legal entity and its operations and administration are independent of those of the university or college, the school or department will be registered rather than the university or college.

(3) A subsidiary of a business corporation, rather than the parent corporation, will be registered as a research facility unless the subsidiary is under such direct control of the parent corporation that the Secretary determines that it is necessary that the parent corporation be registered to effectuate the purposes of the Act.

(b) *Acknowledgment of regulations and standards.* APHIS will supply a copy of the regulations and standards in this subchapter with each registration form. The research facility shall acknowledge receipt of and shall agree to comply with the regulations and standards by signing a form provided for this purpose by APHIS, and by filing it with the AC Regional Director.

(c) *Notification of change of operation.*

(1) A research facility shall notify the AC Regional Director by certified mail of any change in the name, address, or ownership, or other change in operations affecting its status as a research facility, within 10 days after making such change.

(2) A research facility which has not used, handled, or transported animals for a period of at least 2 years may be placed in an inactive status by making a written request to the AC Regional Director. A research facility shall file an annual report of its status (active or inactive). A research facility shall notify the AC Regional Director in writing at least 10 days before using, handling, or transporting animals again after being in an inactive status.

(3) A research facility which goes out of business or which ceases to function as a research facility, or which changes its method of operation so that it no longer uses, handles, or transports animals, and which does not plan to use, handle, or transport animals at any time in the future, may have its registration canceled by making a written request to the AC Regional Director. The research facility is responsible for reregistering and demonstrating its compliance with the Act and regulations should it start using, handling, or transporting animals at any time after its registration is canceled.

(d) No research facility shall interfere with, threaten, abuse (including verbally abuse), or harass any APHIS official who is in the course of carrying out his or her duties.

[54 FR 36147, Aug. 31, 1989, as amended at 63 FR 62926, Nov. 10, 1998; 69 FR 42101, July 14, 2004]

§ 2.31 - Institutional Animal Care and Use Committee (IACUC).

(a) The Chief Executive Officer of the research facility shall appoint an Institutional Animal Care and Use Committee (IACUC), qualified through the experience and expertise of its members to assess the research facility's animal program, facilities, and procedures. Except as specifically authorized by law or these regulations, nothing in this part shall be deemed to permit the Committee or IACUC to prescribe methods or set standards for the design, performance, or conduct of actual research or experimentation by a research facility.

(b) *IACUC membership.*

(1) The members of each Committee shall be appointed by the Chief Executive Officer of the research facility;

(2) The Committee shall be composed of a Chairman and at least two additional members;

(3) Of the members of the Committee:

(i) At least one shall be a Doctor of Veterinary Medicine, with training or experience in laboratory animal science and medicine, who has direct or delegated program responsibility for activities involving animals at the research facility;

(ii) At least one shall not be affiliated in any way with the facility other than as a member of the Committee, and shall not be a member of the immediate family of a person who is affiliated with the facility. The Secretary intends that such person will provide representation for general community interests in the proper care and treatment of animals;

(4) If the Committee consists of more than three members, not more than three members shall be from the same administrative unit of the facility.

(c) *IACUC functions.* With respect to activities involving animals, the IACUC, as an agent of the research facility, shall:

(1) Review, at least once every six months, the research facility's program for humane care and use of animals, using title 9, chapter I, subchapter A – Animal Welfare, as a basis for evaluation;

(2) Inspect, at least once every six months, all of the research facility's animal facilities, including animal study areas, using title 9, chapter I, subchapter A-Animal Welfare, as a basis for evaluation; *Provided, however,* That animal areas containing free-living wild animals in their natural habitat need not be included in such inspection;

(3) Prepare reports of its evaluations conducted as required by paragraphs (c)(1) and (2) of this section, and submit the reports to the Institutional Official of the research facility; *Provided, however,* That the IACUC may determine the best means of conducting evaluations of the research facility's programs and facilities; and *Provided, further,* That no Committee member wishing to participate in any evaluation conducted under this subpart may be excluded. The IACUC may use subcommittees composed of at least two Committee members and may invite ad hoc consultants to assist in conducting the evaluations, however, the IACUC remains responsible for the evaluations and reports as required by the Act and regulations. The reports shall be reviewed and signed by a majority of the IACUC members and must include any minority views. The reports shall be updated at least once every six months upon completion of the required semiannual evaluations and shall be maintained by the research facility and made available to APHIS and to officials of funding Federal agencies for inspection and copying upon request. The reports must contain a description of the nature and extent of the research facility's adherence to this subchapter, must identify specifically any departures from the provisions of title 9, chapter I, subchapter A – Animal Welfare, and must state the reasons for each departure. The reports must distinguish significant deficiencies from minor deficiencies. A significant deficiency is one which, with reference to Subchapter A, and, in the judgment of the IACUC and the Institutional Official, is or may be a threat to the health or safety of the animals. If program or facility deficiencies are noted, the reports must contain a reasonable and specific plan and schedule with dates for correcting each deficiency. Any failure to adhere to the plan and schedule that results in a significant deficiency remaining uncorrected shall be reported in writing within 15 business days by the IACUC, through the Institutional Official, to APHIS and any Federal agency funding that activity;

(4) Review, and, if warranted, investigate concerns involving the care and use of animals at the research facility resulting from public complaints

PART 2
Subpart C

received and from reports of noncompliance received from laboratory or research facility personnel or employees;

(5) Make recommendations to the Institutional Official regarding any aspect of the research facility's animal program, facilities, or personnel training;

(6) Review and approve, require modifications in (to secure approval), or withhold approval of those components of proposed activities related to the care and use of animals, as specified in paragraph (d) of this section;

(7) Review and approve, require modifications in (to secure approval), or withhold approval of proposed significant changes regarding the care and use of animals in ongoing activities; and

(8) Be authorized to suspend an activity involving animals in accordance with the specifications set forth in paragraph (d)(6) of this section.

(d) *IACUC review of activities involving animals.*

(1) In order to approve proposed activities or proposed significant changes in ongoing activities, the IACUC shall conduct a review of those components of the activities related to the care and use of animals and determine that the proposed activities are in accordance with this subchapter unless acceptable justification for a departure is presented in writing; *Provided, however,* That field studies as defined in part 1 of this subchapter are exempt from this requirement. Further, the IACUC shall determine that the proposed activities or significant changes in ongoing activities meet the following requirements:

(i) Procedures involving animals will avoid or minimize discomfort, distress, and pain to the animals;

(ii) The principal investigator has considered alternatives to procedures that may cause more than momentary or slight pain or distress to the animals, and has provided a written narrative description of the methods and sources, e.g., the Animal Welfare Information Center, used to determine that alternatives were not available;

(iii) The principal investigator has provided written assurance that the activities do not unnecessarily duplicate previous experiments;

(iv) Procedures that may cause more than momentary or slight pain or distress to the animals will:

(A) Be performed with appropriate sedatives, analgesics or anesthetics, unless withholding such agents is justified for scientific reasons, in writing, by the principal investigator and will continue for only the necessary period of time;

(B) Involve, in their planning, consultation with the attending veterinarian or his or her designee;

(C) Not include the use of paralytics without anesthesia;

PART 2
Subpart C

(**v**) Animals that would otherwise experience severe or chronic pain or distress that cannot be relieved will be painlessly euthanized at the end of the procedure or, if appropriate, during the procedure;

(**vi**) The animals' living conditions will be appropriate for their species in accordance with part 3 of this subchapter, and contribute to their health and comfort. The housing, feeding, and nonmedical care of the animals will be directed by the attending veterinarian or other scientist trained and experienced in the proper care, handling, and use of the species being maintained or studied;

(**vii**) Medical care for animals will be available and provided as necessary by a qualified veterinarian;

(**viii**) Personnel conducting procedures on the species being maintained or studied will be appropriately qualified and trained in those procedures;

(**ix**) Activities that involve surgery include appropriate provision for pre-operative and post-operative care of the animals in accordance with established veterinary medical and nursing practices. All survival surgery will be performed using aseptic procedures, including surgical gloves, masks, sterile instruments, and aseptic techniques. Major operative procedures on non-rodents will be conducted only in facilities intended for that purpose which shall be operated and maintained under aseptic conditions. Non-major operative procedures and all surgery on rodents do not require a dedicated facility, but must be performed using aseptic procedures. Operative procedures conducted at field sites need not be performed in dedicated facilities, but must be performed using aseptic procedures;

(**x**) No animal will be used in more than one major operative procedure from which it is allowed to recover, unless:

(**A**) Justified for scientific reasons by the principal investigator, in writing;

(**B**) Required as routine veterinary procedure or to protect the health or well-being of the animal as determined by the attending veterinarian; or

(**C**) In other special circumstances as determined by the Administrator on an individual basis. Written requests and supporting data should be sent to the Animal and Plant Health Inspection Service, Animal Care, 4700 River Road, Unit 84, Riverdale, Maryland 20737-1234;

(**xi**) Methods of euthanasia used must be in accordance with the definition of the term set forth in 9 CFR part 1, § 1.1 of this subchapter, unless a deviation is justified for scientific reasons, in writing, by the investigator.

(**2**) Prior to IACUC review, each member of the Committee shall be provided with a list of proposed activities to be reviewed. Written

PART 2
Subpart C

descriptions of all proposed activities that involve the care and use of animals shall be available to all IACUC members, and any member of the IACUC may obtain, upon request, full Committee review of those activities. If full Committee review is not requested, at least one member of the IACUC, designated by the chairman and qualified to conduct the review, shall review those activities, and shall have the authority to approve, require modifications in (to secure approval), or request full Committee review of any of those activities. If full Committee review is requested for a proposed activity, approval of that activity may be granted only after review, at a convened meeting of a quorum of the IACUC, and with the approval vote of a majority of the quorum present. No member may participate in the IACUC review or approval of an activity in which that member has a conflicting interest (e.g., is personally involved in the activity), except to provide information requested by the IACUC, nor may a member who has a conflicting interest contribute to the constitution of a quorum;

(3) The IACUC may invite consultants to assist in the review of complex issues arising out of its review of proposed activities. Consultants may not approve or withhold approval of an activity, and may not vote with the IACUC unless they are also members of the IACUC;

(4) The IACUC shall notify principal investigators and the research facility in writing of its decision to approve or withhold approval of those activities related to the care and use of animals, or of modifications required to secure IACUC approval. If the IACUC decides to withhold approval of an activity, it shall include in its written notification a statement of the reasons for its decision and give the principal investigator an opportunity to respond in person or in writing. The IACUC may reconsider its decision, with documentation in Committee minutes, in light of the information provided by the principal investigator;

PART 2
Subpart C

(5) The IACUC shall conduct continuing reviews of activities covered by this subchapter at appropriate intervals as determined by the IACUC, but not less than annually;

(6) The IACUC may suspend an activity that it previously approved if it determines that the activity is not being conducted in accordance with the description of that activity provided by the principal investigator and approved by the Committee. The IACUC may suspend an activity only after review of the matter at a convened meeting of a quorum of the IACUC and with the suspension vote of a majority of the quorum present;

(7) If the IACUC suspends an activity involving animals, the Institutional Official, in consultation with the IACUC, shall review the reasons for suspension, take appropriate corrective action, and report that action with a full explanation to APHIS and any Federal agency funding that activity; and

(8) Proposed activities and proposed significant changes in ongoing activities that have been approved by the IACUC may be subject to further appropriate review and approval by officials of the research facility. However, those officials may not approve an activity involving the care and use of animals if it has not been approved by the IACUC.

(e) A proposal to conduct an activity involving animals, or to make a significant change in an ongoing activity involving animals, must contain the following:

(1) Identification of the species and the approximate number of animals to be used;

(2) A rationale for involving animals, and for the appropriateness of the species and numbers of animals to be used;

(3) A complete description of the proposed use of the animals;

(4) A description of procedures designed to assure that discomfort and pain to animals will be limited to that which is unavoidable for the conduct of scientifically valuable research, including provision for the use of analgesic, anesthetic, and tranquilizing drugs where indicated and appropriate to minimize discomfort and pain to animals; and

(5) A description of any euthanasia method to be used.

[54 FR 36147, Aug. 31, 1989, as amended at 59 FR 67611, Dec. 30, 1994; 63 FR 62926, Nov. 10, 1998]

§ 2.32 - Personnel qualifications.

(a) It shall be the responsibility of the research facility to ensure that all scientists, research technicians, animal technicians, and other personnel involved in animal care, treatment, and use are qualified to perform their duties. This responsibility shall be fulfilled in part through the provision of training and instruction to those personnel.

(b) Training and instruction shall be made available, and the qualifications of personnel reviewed, with sufficient frequency to fulfill the research facility's responsibilities under this section and § 2.31.

(c) Training and instruction of personnel must include guidance in at least the following areas:

(1) Humane methods of animal maintenance and experimentation, including:

(i) The basic needs of each species of animal;

(ii) Proper handling and care for the various species of animals used by the facility;

(iii) Proper pre-procedural and post-procedural care of animals; and

(iv) Aseptic surgical methods and procedures;

(2) The concept, availability, and use of research or testing methods that limit the use of animals or minimize animal distress;

(3) Proper use of anesthetics, analgesics, and tranquilizers for any species of animals used by the facility;

(4) Methods whereby deficiencies in animal care and treatment are reported, including deficiencies in animal care and treatment reported by any employee of the facility. No facility employee, Committee member, or laboratory personnel shall be discriminated against or be subject to any reprisal for reporting violations of any regulation or standards under the Act;

(5) Utilization of services (e.g., National Agricultural Library, National Library of Medicine) available to provide information:

(i) On appropriate methods of animal care and use;

(ii) On alternatives to the use of live animals in research;

(iii) That could prevent unintended and unnecessary duplication of research involving animals; and

(iv) Regarding the intent and requirements of the Act.

§ 2.33 - Attending veterinarian and adequate veterinary care.

(a) Each research facility shall have an attending veterinarian who shall provide adequate veterinary care to its animals in compliance with this section:

(1) Each research facility shall employ an attending veterinarian under formal arrangements. In the case of a part-time attending veterinarian or consultant arrangements, the formal arrangements shall include a written program of veterinary care and regularly scheduled visits to the research facility;

(2) Each research facility shall assure that the attending veterinarian has appropriate authority to ensure the provision of adequate veterinary care and to oversee the adequacy of other aspects of animal care and use; and

(3) The attending veterinarian shall be a voting member of the IACUC; *Provided, however,* That a research facility with more than one Doctor of Veterinary Medicine (DVM) may appoint to the IACUC another DVM with delegated program responsibility for activities involving animals at the research facility.

(b) Each research facility shall establish and maintain programs of adequate veterinary care that include:

(1) The availability of appropriate facilities, personnel, equipment, and services to comply with the provisions of this subchapter;

(2) The use of appropriate methods to prevent, control, diagnose, and treat diseases and injuries, and the availability of emergency, weekend, and holiday care;

(3) Daily observation of all animals to assess their health and well-being; *Provided, however,* That daily observation of animals may be accomplished by someone other than the attending veterinarian; and *Provided, further,* That a mechanism of direct and frequent communication is required so that timely and accurate information on problems of animal health, behavior, and well-being is conveyed to the attending veterinarian;

(4) Guidance to principal investigators and other personnel involved in the care and use of animals regarding handling, immobilization, anesthesia, analgesia, tranquilization, and euthanasia; and

(5) Adequate pre-procedural and post-procedural care in accordance with current established veterinary medical and nursing procedures.

§ 2.34 - [Reserved]

§ 2.35 - Recordkeeping requirements.

(a) The research facility shall maintain the following IACUC records:

(1) Minutes of IACUC meetings, including records of attendance, activities of the Committee, and Committee deliberations;

(2) Records of proposed activities involving animals and proposed significant changes in activities involving animals, and whether IACUC approval was given or withheld; and

(3) Records of semiannual IACUC reports and recommendations (including minority views), prepared in accordance with the requirements of § 2.31(c)(3) of this subpart, and forwarded to the Institutional Official.

(b) Every research facility shall make, keep, and maintain records or forms which fully and correctly disclose the following information concerning each live dog or cat purchased or otherwise acquired, owned, held, or otherwise in their possession or under their control, transported, euthanized, sold, or otherwise disposed of by the research facility. The records shall include any offspring born of any animal while in the research facility's possession or under its control:

(1) The name and address of the person from whom a dog or cat was purchased or otherwise acquired, whether or not the person is required to be licensed or registered under the Act;

(2) The USDA license or registration number of the person if he or she is licensed or registered under the Act;

(3) The vehicle license number and State, and the driver's license number (or photographic identification card for non-drivers issued by a State) and State of the person, if he or she is not licensed or registered under the Act;

(4) The date of acquisition of each dog or cat;

(5) The official USDA tag number or tattoo assigned to each dog or cat under § 2.38(g) of this subpart;

(6) A description of each dog or cat which shall include:

(i) The species and breed or type of animal;

(ii) The sex;

(iii) The date of birth or approximate age; and

(iv) The color and any distinctive markings;

(7) Any identification number or mark assigned to each dog or cat by the research facility;

(8) If dogs or cats are acquired from any person not licensed or registered under the Act and not a pound or shelter, the research facility must obtain a certification that the animals were born and raised on the person's premises and that the person has sold fewer than 25 dogs and/or cats that year.

(c) In addition to the information required to be kept and maintained by every research facility concerning each live dog or cat under paragraph (a) of this section, every research facility transporting, selling, or otherwise disposing of any live dog or cat to another person, shall make and maintain records or forms which fully and correctly disclose the following information:

(1) The name and address of the person to whom a live dog or cat is transported, sold, or otherwise disposed of;

(2) The date of transportation, sale, euthanasia, or other disposition of the animal; and

PART 2
Subpart C

(3) The method of transportation, including the name of the initial carrier or intermediate handler, or if a privately owned vehicle is used to transport the dog or cat, the name of the owner of the privately owned vehicle.

(d)(1) The USDA Interstate and International Certificate of Health Examination for Small Animals (APHIS Form 7001) and Record of Acquisition and Dogs and Cats on Hand (APHIS Form 7005) are forms which may be used by research facilities to keep and maintain the information required by paragraph (b) of this section.

(2) The USDA Interstate and International Certificate of Health Examination for Small Animals (APHIS Form 7001) and Record of Disposition of Dogs and Cats (APHIS Form 7006) are forms which may be used by research facilities to keep and maintain the information required by paragraph (c) of this section.

(e) One copy of the record containing the information required by paragraphs (b) and (c) of this section shall accompany each shipment of any live dog or cat sold or otherwise disposed of by a research facility; *Provided, however,* That, except as provided in § 2.133 of this part, information that indicates the source and date of acquisition of any dog or cat need not appear on the copy of the record accompanying the shipment. One copy of the record containing the information required by paragraphs (b) and (c) of this section shall be retained by the research facility.

(f) All records and reports shall be maintained for at least three years. Records that relate directly to proposed activities and proposed significant changes in ongoing activities reviewed and approved by the IACUC shall be maintained for the duration of the activity and for an additional three years after completion of the activity. All records shall be available for inspection and copying by authorized APHIS or funding Federal agency representatives at reasonable times. APHIS inspectors will maintain the confidentiality of the information and will not remove the materials from the research facilities' premises unless there has been an alleged violation, they are needed to investigate a possible violation, or for other enforcement purposes. Release of any such materials, including reports, summaries, and photographs that contain trade secrets or commercial or financial information that is privileged or confidential will be governed by applicable sections of the Freedom of Information Act. Whenever the Administrator notifies a research facility in writing that specified records shall be retained pending completion of an investigation or proceeding under the Act, the research facility shall hold those records until their disposition is authorized in writing by the Administrator.

(Approved by the Office of Management and Budget under control number 0579-0254)

[54 FR 36147, Aug. 31, 1989, as amended at 58 FR 39129, July 22, 1993; 60 FR 13895, Mar. 15, 1995; 69 FR 42101, July 14, 2004]

PART 2
Subpart C

§ 2.36 - Annual report.

(a) The reporting facility shall be that segment of the research facility, or that department, agency, or instrumentality of the United States, that uses or intends to use live animals in research, tests, experiments, or for teaching. Each reporting facility shall submit an annual report to the AC Regional Director for the State where the facility is located on or before December 1 of each calendar year. The report shall be signed and certified by the CEO or Institutional Official, and shall cover the previous Federal fiscal year.

(b) The annual report shall:

(1) Assure that professionally acceptable standards governing the care, treatment, and use of animals, including appropriate use of anesthetic, analgesic, and tranquilizing drugs, prior to, during, and following actual research, teaching, testing, surgery, or experimentation were followed by the research facility;

(2) Assure that each principal investigator has considered alternatives to painful procedures;

(3) Assure that the facility is adhering to the standards and regulations under the Act, and that it has required that exceptions to the standards and regulations be specified and explained by the principal investigator and approved by the IACUC. A summary of all such exceptions must be attached to the facility's annual report. In addition to identifying the IACUC-approved exceptions, this summary must include a brief explanation of the exceptions, as well as the species and number of animals affected;

(4) State the location of all facilities where animals were housed or used in actual research, testing, teaching, or experimentation, or held for these purposes;

(5) State the common names and the numbers of animals upon which teaching, research, experiments, or tests were conducted involving no pain, distress, or use of pain-relieving drugs. Routine procedures (e.g., injections, tattooing, blood sampling) should be reported with this group;

(6) State the common names and the numbers of animals upon which experiments, teaching, research, surgery, or tests were conducted involving accompanying pain or distress to the animals and for which appropriate anesthetic, analgesic, or tranquilizing drugs were used;

(7) State the common names and the numbers of animals upon which teaching, experiments, research, surgery, or tests were conducted involving accompanying pain or distress to the animals and for which the use of appropriate anesthetic, analgesic, or tranquilizing drugs would have adversely affected the procedures, results, or interpretation of the teaching, research, experiments, surgery, or tests. An explanation of the procedures producing pain or distress in these animals and the reasons such drugs were not used shall be attached to the annual report;

(8) State the common names and the numbers of animals being bred, conditioned, or held for use in teaching, testing, experiments, research, or surgery but not yet used for such purposes.

[54 FR 36147, Aug. 31, 1989, as amended at 63 FR 62926, Nov. 10, 1998]

§ 2.37 - Federal research facilities.

Each Federal research facility shall establish an Institutional Animal Care and Use Committee which shall have the same composition, duties, and responsibilities required of nonfederal research facilities by § 2.31 with the following exceptions:

PART 2
Subpart C

(a) The Committee shall report deficiencies to the head of the Federal agency conducting the research rather than to APHIS; and

(b) The head of the Federal agency conducting the research shall be responsible for all corrective action to be taken at the facility and for the granting of all exceptions to inspection protocol.

§ 2.38 - Miscellaneous.

(a) *Information as to business: furnishing of same by research facilities.* Each research facility shall furnish to any APHIS official any information concerning the business of the research facility which the APHIS official may request in connection with the enforcement of the provisions of the Act, the regulations, and the standards in this subchapter. The information shall be furnished within a reasonable time and as may be specified in the request for information.

(b) *Access and inspection of records and property.*

    (1) Each research facility shall, during business hours, allow APHIS officials:

        (i) To enter its place of business;

        (ii) To examine records required to be kept by the Act and the regulations in this part;

        (iii) To make copies of the records;

        (iv) To inspect the facilities, property, and animals, as the APHIS officials consider necessary to enforce the provisions of the Act, the regulations, and the standards in this subchapter; and

        (v) To document, by the taking of photographs and other means, conditions and areas of noncompliance.

    (2) The use of a room, table or other facilities necessary for the proper examination of the records and for inspection of the property or animals shall be extended to APHIS officials by the research facility.

(c) *Publication of names of research facilities subject to the provisions of this part.* APHIS will publish lists of research facilities registered in accordance with the provisions of this subpart in the FEDERAL REGISTER. The lists may be obtained upon request from the AC Regional Director.

(d) *Inspection for missing animals.* Each research facility shall allow, upon request and during business hours, police or officers of other law enforcement agencies with general law enforcement authority (not those agencies whose duties are limited to enforcement of local animal regulations) to enter its place of business to inspect animals and records for the purpose of seeking animals that are missing, under the following conditions:

    (1) The police or other law officer shall furnish to the research facility a written description of the missing animal and the name and address of its owner before making a search;

PART 2
Subpart C

(2) The police or other law officer shall abide by all security measures required by the research facility to prevent the spread of disease, including the use of sterile clothing, footwear, and masks where required, or to prevent the escape of an animal.

(e) *Confiscation and destruction of animals.*

(1) If an animal being held by a research facility is not being used to carry out research, testing, or experimentation, and is found by an APHIS official to be suffering as a result of the failure of the research facility to comply with any provision of the regulations or the standards set forth in this subchapter, the APHIS official shall make a reasonable effort to notify the research facility of the condition of the animal(s) and request that the condition be corrected and that adequate care be given to alleviate the animal's suffering or distress, or that the animal(s) be destroyed by euthanasia. In the event that the research facility refuses to comply with this request, the APHIS official may confiscate the animal(s) for care, treatment, or disposal as indicated in paragraph (e)(2) of this section, if, in the opinion of the Administrator, the circumstances indicate the animal's health is in danger.

(2) In the event that the APHIS official is unable to locate or notify the research facility as required in this section, the APHIS official shall contact a local police or other law officer to accompany him or her to the premises and shall provide for adequate care when necessary to alleviate the animal's suffering. If, in the opinion of the Administrator, the condition of the animal(s) cannot be corrected by this temporary care, the APHIS official shall confiscate the animal(s).

PART 2
Subpart C

(3) Confiscated animals may be placed, by sale or donation, with other registrants or licensees that comply with the standards and regulations and can provide proper care, or they may be euthanized. The research facility from which the animals were confiscated shall bear all costs incurred in performing the placement or euthanasia activities authorized by this section.

(f) *Handling.*

(1) Handling of all animals shall be done as expeditiously and carefully as possible in a manner that does not cause trauma, overheating, excessive cooling, behavioral stress, physical harm, or unnecessary discomfort.

(2)(i) Physical abuse shall not be used to train, work, or otherwise handle animals.

(ii) Deprivation of food or water shall not be used to train, work, or otherwise handle animals; *Provided, however:* That the short-term withholding of food or water from animals, when specified in an IACUC-approved activity that includes a description of monitoring procedures, is allowed by these regulations.

(g) *Identification of dogs and cats.*

(1) All live dogs or cats, including those from any exempt source, delivered for transportation, transported, purchased or otherwise acquired. sold, or disposed of by a research facility, shall be identified at the time of such delivery for transportation, purchase, sale, disposal, or acquisition in one of the following ways:

(i) By the official tag or tattoo which was affixed to the animal at the time it was acquired by the research facility, as required by this section; or

(ii) By a tag, tattoo, or collar, applied to the live dog or cat by the research facility and which individually identifies the dog or cat by number.

(2) All official tag or tattoo numbers shall be correctly listed in the records of purchase, acquisition, disposal, or sale which shall be maintained in accordance with § 2.35.

(3) Unweaned puppies or kittens need not be individually identified while they are maintained as a litter with their dam in the same primary enclosure, provided the dam has been individually identified.

(4) The official tag shall be made of a durable alloy such as brass, bronze, or steel, or of a durable plastic. Aluminum of a sufficient thickness to assure the tag is durable and legible may also be used. The tag may be circular in shape and not less than 1¼ inches in diameter, or oblong and flat in shape and not less than 2 inches by ¾ inch, and riveted to an acceptable collar.

(5) Each tag shall have the following information embossed or stamped on so that it is easily readable:

(i) The letters "USDA";

(ii) Numbers identifying the State and dealer, exhibitor, or research facility (e.g., 39-AB); and

(iii) Numbers identifying the animal (e.g., 82488).

(6) Official tags shall be serially numbered and shall be applied to dogs or cats in the manner set forth in this section in as close to consecutive numerical order as possible. No tag number shall be used to identify more than one animal or shall be reused within a 5-year period.

(7) Research facilities may obtain, at their own expense, official tags from commercial tag manufacturers.[1] At the time the research facility is registered, the Department will assign identification letters and numbers to be used on the official tags.

(8) Each research facility shall be held accountable for all official tags acquired. In the event an official tag is lost from a dog or cat while in the possession of a research facility, the facility shall make a diligent effort to locate and reapply the tag to the proper animal. If the lost tag is not located,

PART 2
Subpart C

---

1 *A list of the commercial manufacturers who produce these tags and are known to the Department may be obtained from the AC Regional Director. Any manufacturer who desires to be included in the list should notify the Administrator.*

the research facility shall affix another official tag to the animal in the manner prescribed in this section and record the tag number on the official records.

(9) When a dog or cat wearing or identified by an official tag arrives at a research facility, the facility may continue to use that tag to identify the dog or cat or the tag may be replaced as indicated in paragraph (g)(1) of this section. All tags removed by a research facility shall be retained and disposed of as indicated in this section.

(10) Where a dog or cat to which is affixed or which is identified by an official tag is euthanized, or dies from other causes, the research facility shall remove and retain the tag for the required period, as set forth in paragraph (g)(11) of this section.

(11) All official tags removed and retained by a research facility shall be held until called for by an APHIS official or for a period of 1 year.

(12) When official tags are removed from animals for disposal, the tags must be disposed of so as to preclude their reuse for animal identification. No animal identification number shall be used within any 5-year period following its previous use.

(h) *Health certification.*

(1) No research facility, including a Federal research facility, shall deliver to any intermediate handler or carrier for transportation, in commerce, or shall transport in commerce any dog, cat, or nonhuman primate unless the dog, cat, or nonhuman primate is accompanied by a health certificate executed and issued by a licensed veterinarian. The health certificate shall state that:

PART 2
Subpart C

(i) The licensed veterinarian inspected the dog, cat, or nonhuman primate on a specified date which shall not be more than 10 days prior to the delivery of the dog, cat, or nonhuman primate for transportation; and

(ii) When so inspected, the dog, cat, or nonhuman primate appeared to the licensed veterinarian to be free of any infectious disease or physical abnormality which would endanger the animal(s) or other animals or endanger public health.

(2) The Secretary may provide exceptions to the health certification requirement on an individual basis for animals shipped to a research facility for purposes of research, testing, or experimentation when the research facility requires animals not eligible for certification. Requests should be addressed to the Animal and Plant Health Inspection Service, Animal Care, 4700 River Road, Unit 84, Riverdale, Maryland 20737-1234.

(3) The U.S. Interstate and International Certificate of Health Examination for Small Animals (APHIS Form 7001) may be used for health certification by a licensed veterinarian as required by this section.

(i) *Holding of animals.* If any research facility obtains prior approval of the AC Regional Director, it may arrange to have another person hold animals: *Provided,* That:

(1) The other person agrees, in writing, to comply with the regulations in this part and the standards in part 3 of this subchapter, and to allow inspection of the premises by an APHIS official during business hours;

(2) The animals remain under the total control and responsibility of the research facility; and

(3) The Institutional Official agrees, in writing, that the other person or premises is a recognized animal site under its research facility registration. APHIS Form 7009 shall be used for approval.

(4) The other person or premises must either be directly included in the research facility's contingency plan required under paragraph (l) of this section or must develop its own contingency plan in accordance with paragraph (l) of this section.

(j) *Holding period.* Research facilities that obtain dogs and cats from sources other than dealers, exhibitors, and exempt persons shall hold the animals for 5 full days, not including the day of acquisition, after acquiring the animal, excluding time in transit, before they may be used by the facility. Research facilities shall comply with the identification of animals requirements set forth in § 2.38(g) during this period.

(k) *Compliance with standards and prohibitions.*

(1) Each research facility shall comply in all respects with the regulations set forth in subpart C of this part and the standards set forth in part 3 of this subchapter for the humane handling, care, treatment, housing, and transportation of animals; *Provided, however,* That exceptions to the standards in part 3 and the provisions of subpart C of this part may be made only when such exceptions are specified and justified in the proposal to conduct the activity and are approved by the IACUC.

(2) No person shall obtain live dogs or cats by use of false pretenses, misrepresentation, or deception.

(3) No person shall acquire, buy, sell, exhibit, use for research, transport, or offer for transportation, any stolen animal.

(4) Each research facility shall comply with the regulations set forth in § 2.133 of subpart I of this part.

(l) *Contingency planning.*

(1) Research facilities must develop, document, and follow an appropriate plan to provide for the humane handling, treatment, transportation, housing, and care of their animals in the event of an emergency or disaster (one which could reasonably be anticipated and expected to be detrimental to the good health and well-being of the animals in their possession). Such contingency plans must:

PART 2
Subpart C

(i) Identify situations the facility might experience that would trigger the need for the measures identified in a contingency plan to be put into action including, but not limited to, emergencies such as electrical outages, faulty HVAC systems, fires, and animal escapes, as well as natural disasters the facility is most likely to experience.

(ii) Outline specific tasks required to be carried out in response to the identified emergencies or disasters including, but not limited to, detailed animal evacuation instructions or shelter-in-place instructions and provisions for providing backup sources of food and water as well as sanitation, ventilation, bedding, veterinary care, etc.;

(iii) Identify a chain of command and who (by name or by position title) will be responsible for fulfilling these tasks; and

(iv) Address how response and recovery will be handled in terms of materials, resources, and training needed.

(2) For current registrants, the contingency plan must be in place by July 29, 2013. For research facilities registered after this date, the contingency plan must be in place prior to conducting regulated activities. The plan must be reviewed by the research facility on at least an annual basis to ensure that it adequately addresses the criteria listed in paragraph (l)(1) of this section. Each registrant must maintain documentation of their annual reviews, including documenting any amendments or changes made to their plan since the previous year's review, such as changes made as a result of recently predicted, but historically unforeseen, circumstances (e.g., weather extremes). Contingency plans, as well as all annual review documentation and training records, must be made available to APHIS and any funding Federal agency representatives upon request. Facilities maintaining or otherwise handling marine mammals in captivity must also comply with the requirements of § 3.101(b) of this subchapter.

(3) The facility must provide and document participation in and successful completion of training for its personnel regarding their roles and responsibilities as outlined in the plan. For current registrants, training of facility personnel must be completed by September 27, 2013; for research facilities registered after July 29, 2013, training of facility personnel must be completed within 60 days of the facility putting its contingency plan in place. Employees hired 30 days or more before the contingency plan is put in place must also be trained by that date. For employees hired less than 30 days before that date or after that date, training must be conducted within 30 days of their start date. Any changes to the plan as a result of the annual review must be communicated to employees through training which must be conducted within 30 days of making the changes.

PART 2
Subpart C

[54 FR 36147, Aug. 31, 1989, as amended at 58 FR 39129, July 22, 1993; 59 FR 67612, Dec. 30, 1994; 60 FR 13895, Mar. 15, 1995; 63 FR 62926, Nov. 10, 1998; 69 FR 42101, July 14, 2004; 77 FR 76823, Dec. 31, 2012]

Effective Date Note: *At 78 FR 46255, July 31, 2013, in § 2.38, paragraph (l) was stayed indefinitely, effective July 31, 2013.*

PART 2
Subpart C

Subpart D – Attending Veterinarian and Adequate Veterinary Care

§ 2.40 - Attending veterinarian and adequate veterinary care (dealers and exhibitors).

(a) Each dealer or exhibitor shall have an attending veterinarian who shall provide adequate veterinary care to its animals in compliance with this section.

(1) Each dealer and exhibitor shall employ an attending veterinarian under formal arrangements. In the case of a part-time attending veterinarian or consultant arrangements, the formal arrangements shall include a written program of veterinary care and regularly scheduled visits to the premises of the dealer or exhibitor; and

(2) Each dealer and exhibitor shall assure that the attending veterinarian has appropriate authority to ensure the provision of adequate veterinary care and to oversee the adequacy of other aspects of animal care and use.

(b) Each dealer or exhibitor shall establish and maintain programs of adequate veterinary care that include:

(1) The availability of appropriate facilities, personnel, equipment, and services to comply with the provisions of this subchapter;

(2) The use of appropriate methods to prevent, control, diagnose, and treat diseases and injuries, and the availability of emergency, weekend, and holiday care;

(3) Daily observation of all animals to assess their health and well-being; *Provided, however,* That daily observation of animals may be accomplished by someone other than the attending veterinarian; and *Provided, further,* That a mechanism of direct and frequent communication is required so that timely and accurate information on problems of animal health, behavior, and well-being is conveyed to the attending veterinarian;

(4) Adequate guidance to personnel involved in the care and use of animals regarding handling, immobilization, anesthesia, analgesia, tranquilization, and euthanasia; and

PART 2
Subpart D

(5) Adequate pre-procedural and post-procedural care in accordance with established veterinary medical and nursing procedures.

Subpart E – Identification of Animals

§ 2.50 - Time and method of identification.

(a) A class "A" dealer (breeder) shall identify all live dogs and cats on the premises as follows:

(1) All live dogs and cats held on the premises, purchased, or otherwise acquired, sold or otherwise disposed of, or removed from the premises for delivery to a research facility or exhibitor or to another dealer, or for sale, through an auction sale or to any person for use as a pet, shall be identified by an official tag of the type described in § 2.51 affixed to the animal's neck by means of a collar made of material generally considered acceptable to pet owners as a means of identifying their pet dogs or cats[2], or shall be identified by a distinctive and legible tattoo marking acceptable to and approved by the Administrator.

(2) Live puppies or kittens, less than 16 weeks of age, shall be identified by:

(i) An official tag as described in § 2.51;

(ii) A distinctive and legible tattoo marking approved by the Administrator; or

(iii) A plastic-type collar acceptable to the Administrator which has legibly placed thereon the information required for an official tag pursuant to § 2.51.

(b) A class "B" dealer shall identify all live dogs and cats under his or her control or on his or her premises as follows:

(1) When live dogs or cats are held, purchased, or otherwise acquired, they shall be immediately identified:

(i) By affixing to the animal's neck an official tag as set forth in § 2.51 by means of a collar made of material generally acceptable to pet owners as a means of identifying their pet dogs or cats[3]; or

(ii) By a distinctive and legible tattoo marking approved by the Administrator.

(2) If any live dog or cat is already identified by an official tag or tattoo which has been applied by another dealer or exhibitor, the dealer or exhibitor who purchases or otherwise acquires the animal may continue identifying the dog or cat by the previous identification number, or may replace the previous tag with his own official tag or approved tattoo. In either case, the class B dealer or class C exhibitor shall correctly list all old and new

PART 2
Subpart E

---

2 *In general, well fitted collars made of leather or plastic will be acceptable under this provision. The use of certain types of chains presently used by some dealers may also be deemed acceptable. APHIS will determine the acceptability of a material proposed for usage as collars from the standpoint of humane considerations on an individual basis in consultation with the dealer or exhibitor involved. The use of materials such as wire, elastic, or sharp metal that might cause discomfort or injury to the dogs or cats is not acceptable.*

3 *See footnote 2 in § 2.50(a)(1).*

official tag numbers or tattoos in his or her records of purchase which shall be maintained in accordance with §§ 2.75 and 2.77. Any new official tag or tattoo number shall be used on all records of any subsequent sales by the dealer or exhibitor, of any dog or cat.

(3) Live puppies or kittens less than 16 weeks of age, shall be identified by:

(i) An official tag as described in § 2.51;

(ii) A distinctive and legible tattoo marking approved by the Administrator; or

(iii) A plastic-type collar acceptable to the Administrator which has legibly placed thereon the information required for an official tag pursuant to § 2.51.

(4) When any dealer has made a reasonable effort to affix an official tag to a cat, as set forth in paragraphs (a) and (b) of this section, and has been unable to do so, or when the cat exhibits serious distress from the attachment of a collar and tag, the dealer shall attach the collar and tag to the door of the primary enclosure containing the cat and take measures adequate to maintain the identity of the cat in relation to the tag. Each primary enclosure shall contain no more than one weaned cat without an affixed collar and official tag, unless the cats are identified by a distinctive and legible tattoo or plastic-type collar approved by the Administrator.

(c) A class "C" exhibitor shall identify all live dogs and cats under his or her control or on his or her premises, whether held, purchased, or otherwise acquired:

(1) As set forth in paragraph (b)(1) or (b)(3) of this section, or

(2) By identifying each dog or cat with:

(i) An official USDA sequentially numbered tag that is kept on the door of the animal's cage or run;

(ii) A record book containing each animal's tag number, a written description of each animal, the data required by § 2.75(a), and a clear photograph of each animal; and

(iii) A duplicate tag that accompanies each dog or cat whenever it leaves the compound or premises.

(d) Unweaned puppies or kittens need not be individually identified as required by paragraphs (a) and (b) of this section while they are maintained as a litter with their dam in the same primary enclosure, provided the dam has been individually identified.

(e)(1) All animals, except dogs and cats, delivered for transportation, transported, purchased, sold, or otherwise acquired or disposed of by any dealer or exhibitor shall be identified by the dealer or exhibitor at the time of delivery for transportation, purchase, sale, acquisition or disposal, as provided for in this paragraph and in records maintained as required in §§ 2.75 and 2.77.

78

(2) When one or more animals, other than dogs or cats, are confined in a primary enclosure, the animal(s) shall be identified by:

(i) A label attached to the primary enclosure which shall bear a description of the animals in the primary enclosure, including:

(A) The number of animals;

(B) The species of the animals;

(C) Any distinctive physical features of the animals; and

(D) Any identifying marks, tattoos, or tags attached to the animals;

(ii) Marking the primary enclosure with a painted or stenciled number which shall be recorded in the records of the dealer or exhibitor together with:

(A) A description of the animal(s);

(B) The species of the animal(s); and

(C) Any distinctive physical features of the animal(s); or

(iii) A tag or tattoo applied to each animal in the primary enclosure by the dealer or exhibitor which individually identifies each animal by description or number.

(3) When any animal, other than a dog or cat, is not confined in a primary enclosure, it shall be identified on a record, as required by § 2.75, which shall accompany the animal at the time it is delivered for transportation, transported, purchased, or sold, and shall be kept and maintained by the dealer or exhibitor as part of his or her records.

§ 2.51 - Form of official tag.

(a) The official tag shall be made of a durable alloy such as brass, bronze, or steel, or of a durable plastic. Aluminum of a sufficient thickness to assure the tag is durable and legible may also be used. The tag shall be one of the following shapes:

(1) Circular in shape and not less than 1¼ inches in diameter, or

(2) Oblong and flat in shape, not less than 2 inches by ¾ inch and riveted to an acceptable collar.

(b) Each tag shall have the following information embossed or stamped on so that it is easily readable:

(1) The letters "USDA";

(2) Numbers identifying the State and dealer, exhibitor, or research facility (e.g., 39-AB); and

(3) Numbers identifying the animal (e.g., 82488).

(c) Official tags shall be serially numbered. No individual dealer or exhibitor shall use any identification tag number more than once within a 5-year period.

PART 2
Subpart E

§ 2.52 - How to obtain tags.

Dealers or exhibitors may obtain, at their own expense, official tags from commercial tag manufacturers.[4] At the time the dealer or exhibitor is issued a license or is registered, the Department will assign identification letters and numbers and inform them of the identification letters and numbers to be used on the official tags.

[54 FR 36147, Aug. 31, 1989, as amended at 63 FR 62927, Nov. 10, 1998]

§ 2.53 - Use of tags.

Official tags obtained by a dealer, exhibitor, or research facility, shall be applied to dogs or cats in the manner set forth in § 2.50 and in as close to consecutive numerical order as possible. No tag number shall be used to identify more than one animal. No number shall be repeated within a 5-year period.

§ 2.54 - Lost tags.

Each dealer or exhibitor shall be held accountable for all official tags acquired. In the event an official tag is lost from a dog or cat while in the possession of a dealer or exhibitor, the dealer or exhibitor shall make a diligent effort to locate and reapply the tag to the proper animal. If the lost tag is not located, the dealer or exhibitor shall affix another official tag to the animal in the manner prescribed in § 2.50, and record the tag number on the official records.

§ 2.55 - Removal and disposal of tags.

(a) Where a dog or cat to which is affixed or which is identified by an official tag is euthanized, or dies from other causes, the dealer or exhibitor shall remove and retain the tag for the required period, as set forth in paragraph (b) of this section.

(b) All official tags removed and retained by a dealer or exhibitor shall be held until called for by an APHIS official or for a period of 1 year.

(c) When official tags are removed from animals for disposal, the tags must be disposed of so as to preclude their reuse for animal identification. No animal identification number shall be used within any 5-year period following its previous use.

PART 2

Subpart E

---

4 A list of the commercial manufacturers who produce these tags and are known to the Department may be obtained from the AC Regional Director. Any manufacturer who desires to be included in the list should notify the Administrator.

Subpart F – Stolen Animals

§ 2.60 - Prohibition on the purchase, sale, use, or transportation of stolen animals.

No person shall buy, sell, exhibit, use for research, transport, or offer for transportation, any stolen animal.

PART 2
Subpart F

Subpart G – Records

§ 2.75 - Records: Dealers and exhibitors.

(a)(1) Each dealer, other than operators of auction sales and brokers to whom animals are consigned, and each exhibitor shall make, keep, and maintain records or forms which fully and correctly disclose the following information concerning each dog or cat purchased or otherwise acquired, owned, held, or otherwise in his or her possession or under his or her control, or which is transported, euthanized, sold, or otherwise disposed of by that dealer or exhibitor. The records shall include any offspring born of any animal while in his or her possession or under his or her control.

 (i) The name and address of the person from whom a dog or cat was purchased or otherwise acquired whether or not the person is required to be licensed or registered under the Act;

 (ii) The USDA license or registration number of the person if he or she is licensed or registered under the Act;

 (iii) The vehicle license number and State, and the driver's license number (or photographic identification card for non-drivers issued by a State) and State of the person, if he or she is not licensed or registered under the Act;

 (iv) The name and address of the person to whom a dog or cat was sold or given and that person's license or registration number if he or she is licensed or registered under the Act;

 (v) The date a dog or cat was acquired or disposed of, including by euthanasia;

 (vi) The official USDA tag number or tattoo assigned to a dog or cat under §§ 2.50 and 2.54;

 (vii) A description of each dog or cat which shall include:

  (A) The species and breed or type;

  (B) The sex;

  (C) The date of birth or approximate age; and

  (D) The color and any distinctive markings;

 (viii) The method of transportation including the name of the initial carrier or intermediate handler or, if a privately owned vehicle is used to transport a dog or cat, the name of the owner of the privately owned vehicle;

 (ix) The date and method of disposition of a dog or cat, e.g., sale, death, euthanasia, or donation.

 (2) Each dealer and exhibitor shall use Record of Acquisition and Dogs and Cats on Hand (APHIS Form 7005) and Record of Disposition of Dogs and Cats (APHIS Form 7006) to make, keep, and maintain the information required by paragraph (a)(1) of this section: *Provided,* that if a dealer or exhibitor who uses a computerized recordkeeping system believes that

PART 2
Subpart G

APHIS Form 7005 and APHIS Form 7006 are unsuitable for him or her to make, keep, and maintain the information required by paragraph (a)(1) of this section, the dealer or exhibitor may request a variance from the requirement to use APHIS Form 7005 and APHIS Form 7006.

(i) The request for a variance must consist of a written statement describing why APHIS Form 7005 and APHIS Form 7006 are unsuitable for the dealer or exhibitor to make, keep, and maintain the information required by paragraph (a)(1) of this section, and a description of the computerized recordkeeping system the person would use in lieu of APHIS Form 7005 and APHIS Form 7006 to make, keep, and maintain the information required by paragraph (a)(1) of this section. APHIS will advise the person as to the disposition of his or her request for a variance from the requirement to use APHIS Form 7005 and APHIS Form 7006.

(ii) A dealer or exhibitor whose request for a variance has been denied may request a hearing in accordance with the applicable rules of practice for the purpose of showing why the request for a variance should not be denied. The denial of the variance shall remain in effect until the final legal decision has been rendered.

(3) The USDA Interstate and International Certificate of Health Examination for Small Animals (APHIS Form 7001) may be used by dealers and exhibitors to make, keep, and maintain the information required by § 2.79.

(4) One copy of the record containing the information required by paragraph (a)(1) of this section shall accompany each shipment of any dog or cat purchased or otherwise acquired by a dealer or exhibitor. One copy of the record containing the information required by paragraph (a)(1) of this section shall accompany each shipment of any dog or cat sold or otherwise disposed of by a dealer or exhibitor: *Provided, however,* that, except as provided in § 2.133(b) of this part for dealers, information that indicates the source and date of acquisition of a dog or cat need not appear on the copy of the record accompanying the shipment. One copy of the record containing the information required by paragraph (a)(1) of this section shall be retained by the dealer or exhibitor.

(b)(1) Every dealer other than operators of auction sales and brokers to whom animals are consigned, and exhibitor shall make, keep, and maintain records or forms which fully and correctly disclose the following information concerning animals other than dogs and cats, purchased or otherwise acquired, owned, held, leased, or otherwise in his or her possession or under his or her control, or which is transported, sold, euthanized, or otherwise disposed of by that dealer or exhibitor. The records shall include any offspring born of any animal while in his or her possession or under his or her control.

PART 2
Subpart G

84

(i) The name and address of the person from whom the animals were purchased or otherwise acquired;

(ii) The USDA license or registration number of the person if he or she is licensed or registered under the Act;

(iii) The vehicle license number and State, and the driver's license number (or photographic identification card for nondrivers issued by a State) and State of the person, if he or she is not licensed or registered under the Act;

(iv) The name and address of the person to whom an animal was sold or given;

(v) The date of purchase, acquisition, sale, or disposal of the animal(s);

(vi) The species of the animal(s); and

(vii) The number of animals in the shipment.

(2) Record of Animals on Hand (other than dogs and cats) (APHIS Form 7019) and Record of Acquisition, Disposition, or Transport of Animals (other than dogs and cats) (APHIS Form 7020) are forms which may be used by dealers and exhibitors to keep and maintain the information required by paragraph (b)(1) of this section concerning animals other than dogs and cats except as provided in § 2.79.

(3) One copy of the record containing the information required by paragraph (b)(1) of this section shall accompany each shipment of any animal(s) other than a dog or cat purchased or otherwise acquired by a dealer or exhibitor. One copy of the record containing the information required by paragraph (b)(1) of this section shall accompany each shipment of any animal other than a dog or cat sold or otherwise disposed of by a dealer or exhibitor; *Provided, however,* That information which indicates the source and date of acquisition of any animal other than a dog or cat need not appear on the copy of the record accompanying the shipment. The dealer or exhibitor shall retain one copy of the record containing the information required by paragraph (b) (1) of this section.

[54 FR 36147, Aug. 31, 1989, as amended at 58 FR 39129, July 22, 1993; 58 FR 45041, Aug. 26, 1993; 60 FR 13895, Mar. 15, 1995; 69 FR 42102, July 14, 2004]

§ 2.76 - Records: Operators of auction sales and brokers.

(a) Every operator of an auction sale or broker shall make, keep, and maintain records or forms which fully and correctly disclose the following information concerning each animal consigned for auction or sold, whether or not a fee or commission is charged:

PART 2
Subpart G

(1) The name and address of the person who owned or consigned the animal(s) for sale;

(2) The name and address of the buyer or consignee who received the animal;

(3) The USDA license or registration number of the person(s) selling, consigning, buying, or receiving the animals if he or she is licensed or registered under the Act;

(4) The vehicle license number and State, and the driver's license number (or photographic identification card for non-drivers issued by a State) and State of the person, if he or she is not licensed or registered under the Act;

(5) The date of the consignment;

(6) The official USDA tag number or tattoo assigned to the animal under §§ 2.50 and 2.54;

(7) A description of the animal which shall include:

   (i) The species and breed or type of animal;

   (ii) The sex of the animal; and

   (iii) The date of birth or approximate age; and

   (iv) The color and any distinctive markings;

(8) The auction sales number or records number assigned to the animal.

(b) One copy of the record containing the information required by paragraph (a) of this section shall be given to the consignor of each animal, one copy of the record shall be given to the purchaser of each animal: *Provided, however,* That information which indicates the source and date of consignment of any animal need not appear on the copy of the record given the purchaser of any animal. One copy of the record containing the information required by paragraph (a) of this section shall be retained by the operator of such auction sale, or broker, for each animal sold by the auction sale or broker.

[54 FR 36147, Aug. 31, 1989, as amended at 69 FR 42102, July 14, 2004]

§ 2.77 - Records: Carriers and intermediate handlers.

(a) In connection with all live animals accepted for shipment on a C.O.D. basis or other arrangement or practice under which the cost of an animal or the transportation of an animal is to be paid and collected upon delivery of the animal to the consignee, the accepting carrier or intermediate handler, if any, shall keep and maintain a copy of the consignor's written guarantee for the payment of transportation charged for any animal not claimed as provided in § 2.80, including, where necessary, both the return transportation charges and an amount sufficient to reimburse the carrier for out-of-pocket expenses incurred for the care, feeding, and storage of the animal. The carrier or intermediate

PART 2

Subpart G

handler at destination shall also keep and maintain a copy of the shipping document containing the time, date, and method of each attempted notification and the final notification to the consignee and the name of the person notifying the consignee, as provided in § 2.80.

(b) In connection with all live dogs, cats, or nonhuman primates delivered for transportation, in commerce, to any carrier or intermediate handler, by any dealer, research facility, exhibitor, operator of an auction sale, broker, or department, agency or instrumentality of the United States or of any state or local government, the accepting carrier or intermediate handler shall keep and maintain a copy of the health certification completed as required by § 2.79, tendered with each live dog, cat, or nonhuman primate.

§ 2.78 - Health certification and identification.

(a) No dealer, exhibitor, operator of an auction sale, broker, or department, agency, or instrumentality of the United States or of any State or local government shall deliver to any intermediate handler or carrier for transportation, in commerce, or shall transport in commerce any dog, cat, or nonhuman primate unless the dog, cat, or nonhuman primate is accompanied by a health certificate executed and issued by a licensed veterinarian. The health certificate shall state that:

(1) The licensed veterinarian inspected the dog, cat, or nonhuman primate on a specified date which shall not be more than 10 days prior to the delivery of the dog, cat, or nonhuman primate for transportation; and

(2) when so inspected, the dog, cat, or nonhuman primate appeared to the licensed veterinarian to be free of any infectious disease or physical abnormality which would endanger the animal(s) or other animals or endanger public health.

(b) The Secretary may provide exceptions to the health certification requirement on an individual basis for animals shipped to a research facility for purposes of research, testing, or experimentation when the research facility requires animals not eligible for certification. Requests should be addressed to the Animal and Plant Health Inspection Service, Animal Care, 4700 River Road, Unit 84, Riverdale, Maryland 20737-1234.

(c) No intermediate handler or carrier to whom any live dog, cat, or nonhuman primate is delivered for transportation by any dealer, research facility, exhibitor, broker, operator of an auction sale, or department, agency, or instrumentality of the United States or any State or local government shall receive a live dog, cat, or nonhuman primate for transportation, in commerce, unless and until it is accompanied by a health certificate issued by a licensed veterinarian in accordance with paragraph (a) of this section, or an exemption issued by the Secretary in accordance with paragraph (b) of this section.

PART 2
Subpart G

(d) The U.S. Interstate and International Certificate of Health Examination for Small Animals (APHIS Form 7001) may be used for health certification by a licensed veterinarian as required by this section.

[54 FR 36147, Aug. 31, 1989, as amended at 59 FR 67612, Dec. 30, 1994; 60 FR 13896, Mar. 15, 1995; 63 FR 62927, Nov. 10, 1998; 69 FR 42102, July 14, 2004]

§ 2.79 - C.O.D. shipments.

(a) No carrier or intermediate handler shall accept any animal for transportation, in commerce, upon any C.O.D. or other basis where any money is to be paid and collected upon delivery of the animal to the consignee, unless the consignor guarantees in writing the payment of all transportation, including any return transportation, if the shipment is unclaimed or the consignee cannot be notified in accordance with paragraphs (b) and (c) of this section, including reimbursing the carrier or intermediate handler for all out-of-pocket expenses incurred for the care, feeding, and storage or housing of the animal.

(b) Any carrier or intermediate handler receiving an animal at a destination on a C.O.D. or other basis any money is to be paid and collected upon delivery of the animal to the consignee shall attempt to notify the consignee at least once every 6 hours for a period of 24 hours after arrival of the animal at the animal holding area of the terminal cargo facility. The carrier or intermediate handler shall record the time, date, and method of each attempted notification and the final notification to the consignee, and the name of the person notifying the consignee, on the shipping document and on the copy of the shipping document accompanying the C.O.D. shipment. If the consignee cannot be notified of the C.O.D. shipment within 24 hours after its arrival, the carrier or intermediate handler shall return the animal to the consignor, or to whomever the consignor has designated, on the next practical available transportation, in accordance with the written agreement required in paragraph (a) of this section and shall notify the consignor. Any carrier or intermediate handler which has notified a consignee of the arrival of a C.O.D. or other shipment of an animal, where any money is to be paid and collected upon delivery of the animal to the consignee, which is not claimed by the consignee within 48 hours from the time of notification, shall return the animal to the consignor, or to whomever the consignor has designated, on the next practical available transportation, in accordance with the written agreement required in paragraph (a) of this section and shall notify the consignor.

(c) It is the responsibility of any carrier or intermediate handler to hold, feed, and care for any animal accepted for transportation, in commerce,

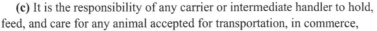

PART 2
Subpart G

88

under a C.O.D. or other arrangement where any money is to be paid and collected upon delivery of the animal until the consignee accepts shipment at destination or until returned to the consignor or his or her designee should the consignee fail to accept delivery of the animal or if the consignee could not be notified as prescribed in paragraph (b) of this section.

(d) Nothing in this section shall be construed as prohibiting any carrier or intermediate handler from requiring any guarantee in addition to that required in paragraph (a) of this section for the payment of the cost of any transportation or out-of-pocket or other incidental expenses incurred in the transportation of any animal.

§ 2.80 - Records, disposition.

(a) No dealer, exhibitor, broker, operator of an auction sale, carrier, or intermediate handler shall, for a period of 1 year, destroy or dispose of, without the consent in writing of the Administrator, any books, records, documents, or other papers required to be kept and maintained under this part.

(b) Unless otherwise specified, the records required to be kept and maintained under this part shall be held for 1 year after an animal is euthanized or disposed of and for any period in excess of one year as necessary to comply with any applicable Federal, State, or local law. Whenever the Administrator notifies a dealer, exhibitor, broker, operator of an auction sale, carrier, or intermediate handler in writing that specified records shall be retained pending completion of an investigation or proceeding under the Act, the dealer, exhibitor, broker, operator of an auction sale, carrier, or intermediate handler shall hold those records until their disposition is authorized by the Administrator.

§ 2.100 - Compliance with standards.

(a) Each dealer, exhibitor, operator of an auction sale, and intermediate handler shall comply in all respects with the regulations set forth in part 2 and the standards set forth in part 3 of this subchapter for the humane handling, care, treatment, housing, and transportation of animals.

(b) Each carrier shall comply in all respects with the regulations in part 2 and the standards in part 3 of this subchapter setting forth the conditions and requirements for the humane transportation of animals in commerce and their handling, care, and treatment in connection therewith.

§ 2.101 - Holding period.

(a) Any live dog or cat acquired by a dealer[5] or exhibitor shall be held by him or her, under his or her supervision and control, for a period of not less than 5 full days, not including the day of acquisition, after acquiring the animal, excluding time in transit: *Provided, however:*

(1) That any live dog or cat acquired by a dealer or exhibitor from any private or contract animal pound or shelter shall be held by that dealer or exhibitor under his or her supervision and control for a period of not less than 10 full days, not including the day of acquisition, after acquiring the animal, excluding time in transit;

(2) Live dogs or cats which have completed a 5-day holding period with another dealer or exhibitor, or a 10-day holding period with another dealer or exhibitor if obtained from a private or contract shelter or pound, may be sold or otherwise disposed of by subsequent dealers or exhibitors after a minimum holding period of 24 hours by each subsequent dealer or exhibitor excluding time in transit;

(3) Any dog or cat suffering from disease, emaciation, or injury may be destroyed by euthanasia prior to the completion of the holding period required by this section; and

(4) Any live dog or cat, 120 days of age or less, that was obtained from the person that bred and raised such dog or cat, may be exempted from the 5-day holding requirement and may be disposed of by dealers or exhibitors after a minimum holding period of 24 hours, excluding time in transit. Each subsequent dealer or exhibitor must also hold each such dog or cat for a 24-hour period excluding time in transit.

(b) During the period in which any dog or cat is being held as required by this section, the dog or cat shall be unloaded from any means of conveyance in which it was received, for food, water, and rest, and shall be handled, cared

---

5 *An operator of an auction sale is not considered to have acquired a dog or cat which is sold through the auction sale.*

for, and treated in accordance with the standards set forth in part 3, subpart A, of this subchapter and § 2.131.

§ 2.102 - Holding facility.

(a) If any dealer or exhibitor obtains the prior approval of the AC Regional Director, he may arrange to have another person hold animals for the required period provided for in paragraph (a) of § 2.101: *Provided,* That:

(1) The other person agrees in writing to comply with the regulations in part 2 and the standards in part 3 of this subchapter and to allow inspection of his premises by an APHIS official during business hours; and

(2) The animals remain under the total control and responsibility of the dealer or exhibitor.

(3) Approval will not be given for a dealer or exhibitor holding a license as set forth in § 2.1 to have animals held for purposes of this section by another licensed dealer or exhibitor. APHIS Form 7009 shall be used for approval.

(4) The other person or premises must either be directly included in the dealer's or exhibitor's contingency plan required under § 2.134 or must develop its own contingency plan in accordance with § 2.134.

(b) If any intermediate handler obtains prior approval of the AC Regional Director, it may arrange to have another person hold animals: *Provided,* That:

(1) The other person agrees in writing to comply with the regulations in part 2 and the standards in part 3 of this subchapter and to allow inspection of the premises by an APHIS official during business hours; and

(2) The animals remain under the total control and responsibility of the research facility or intermediate handler.

(3) The other person or premises must either be directly included in the intermediate handler's contingency plan required under § 2.134 or must develop its own contingency plan in accordance with § 2.134.

[54 FR 36147, Aug. 31, 1989, as amended at 60 FR 13896, Mar. 15, 1995; 63 FR 62927, Nov. 10, 1998; 69 FR 42102, July 14, 2004; 77 FR 76823, Dec. 31, 2012]

Subpart I – Miscellaneous

PART 2
Subpart I

§ 2.125 - Information as to business; furnishing of same by dealers, exhibitors, operators of auction sales, intermediate handlers, and carriers.

Each dealer, exhibitor, operator of an auction sale, intermediate handler, and carrier shall furnish to any APHIS official any information concerning the business of the dealer, exhibitor, operator of an auction sale, intermediate handler or carrier which the APHIS official may request in connection with the enforcement of the provisions of the Act, the regulations and the standards in this subchapter. The information shall be furnished within a reasonable time and as may be specified in the request for information.

§ 2.126 - Access and inspection of records and property; submission of itineraries.

(a) Each dealer, exhibitor, intermediate handler, or carrier, shall, during business hours, allow APHIS officials:

(1) To enter its place of business;

(2) To examine records required to be kept by the Act and the regulations in this part;

(3) To make copies of the records;

(4) To inspect and photograph the facilities, property and animals, as the APHIS officials consider necessary to enforce the provisions of the Act, the regulations and the standards in this subchapter; and

(5) To document, by the taking of photographs and other means, conditions and areas of noncompliance.

(b) The use of a room, table, or other facilities necessary for the proper examination of the records and inspection of the property or animals must be extended to APHIS officials by the dealer, exhibitor, intermediate handler or carrier, and a responsible adult shall be made available to accompany APHIS officials during the inspection process.

(c) Any person who is subject to the Animal Welfare regulations and who intends to exhibit any animal at any location other than the person's approved site (including, but not limited to, circuses, traveling educational exhibits, animal acts, and petting zoos), except for travel that does not extend overnight, shall submit a written itinerary to the AC Regional Director. The itinerary shall be received by the AC Regional Director no fewer than 2 days in advance of any travel and shall contain complete and accurate information concerning the whereabouts of any animal intended for exhibition at any location other than the person's approved site. If the exhibitor accepts an engagement for which travel will begin with less than 48 hours' notice, the exhibitor shall immediately contact the AC Regional Director in writing

PART 2
Subpart I

with the required information. APHIS expects such situations to occur infrequently, and exhibitors who repeatedly provide less than 48 hours' notice will, after notice by APHIS, be subject to increased scrutiny under the Act.

(1) The itinerary shall include the following:

(i) The name of the person who intends to exhibit the animal and transport the animal for exhibition purposes, including any business name and current Act license or registration number and, in the event that any animal is leased, borrowed, loaned, or under some similar arrangement, the name of the person who owns such animal;

(ii) The name, identification number or identifying characteristics, species (common or scientific name), sex and age of each animal; and

(iii) The names, dates, and locations (with addresses) where the animals will travel, be housed, and be exhibited, including all anticipated dates and locations (with addresses) for any stops and layovers that allow or require removal of the animals from the transport enclosures. Unanticipated delays of such length shall be reported to the AC Regional Director the next APHIS business day. APHIS Regional offices are available each weekday, except on Federal holidays, from 8 a.m. to 5 p.m.

(2) The itinerary shall be revised as necessary, and the AC Regional Director shall be notified of any changes. If initial notification of a change due to an emergency is made by a means other than email or facsimile, it shall be followed by written documentation at the earliest possible time. For changes that occur after normal APHIS business hours, the change shall be conveyed to the AC Regional Director no later than the following APHIS business day. APHIS Regional offices are available each weekday, except on Federal holidays, from 8 a.m. to 5 p.m.

(Approved by the Office of Management and Budget under control number 0579-0361)

[54 FR 36147, Aug. 31, 1989, as amended at 69 FR 42102, July 14, 2004; 77 FR 76814, Dec. 31, 2012]

§ 2.127 - Publication of names of persons subject to the provisions of this part.

APHIS will publish lists of persons licensed or registered in accordance with the provisions of this part in the FEDERAL REGISTER. The lists may be obtained upon request from the AC Regional Director.

[54 FR 36147, Aug. 31, 1989, as amended at 63 FR 62927, Nov. 10, 1998]

§ 2.128 - Inspection for missing animals.

Each dealer, exhibitor, intermediate handler and carrier shall allow, upon request and during business hours, police or officers of other law enforcement agencies with general law enforcement authority (not those agencies whose duties are limited to enforcement of local animal regulations) to enter his or her place of business to inspect animals and records for the purpose of seeking animals that are missing, under the following conditions:

PART 2
Subpart I

(a) The police or other law officer shall furnish to the dealer, exhibitor, intermediate handler or carrier a written description of the missing animal and the name and address of its owner before making a search.

(b) The police or other law officer shall abide by all security measures required by the dealer, exhibitor, intermediate handler or carrier to prevent the spread of disease, including the use of sterile clothing, footwear, and masks where required, or to prevent the escape of an animal.

§ 2.129 - Confiscation and destruction of animals.

(a) If an animal being held by a dealer, exhibitor, intermediate handler, or by a carrier is found by an APHIS official to be suffering as a result of the failure of the dealer, exhibitor, intermediate handler, or carrier to comply with any provision of the regulations or the standards set forth in this subchapter, the APHIS official shall make a reasonable effort to notify the dealer, exhibitor, intermediate handler, or carrier of the condition of the animal(s) and request that the condition be corrected and that adequate care be given to alleviate the animal's suffering or distress, or that the animal(s) be destroyed by euthanasia. In the event that the dealer, exhibitor, intermediate handler, or carrier refuses to comply with this request, the APHIS official may confiscate the animal(s) for care, treatment, or disposal as indicated in paragraph (b) of this section, if, in the opinion of the Administrator, the circumstances indicate the animal's health is in danger.

(b) In the event that the APHIS official is unable to locate or notify the dealer, exhibitor, intermediate handler, or carrier as required in this section, the APHIS official shall contact a local police or other law officer to accompany him to the premises and shall provide for adequate care when necessary to alleviate the animal's suffering. If in the opinion of the Administrator, the condition of the animal(s) cannot be corrected by this temporary care, the APHIS official shall confiscate the animals.

(c) Confiscated animals may be:

(1) Placed, by sale or donation, with other licensees or registrants that comply with the standards and regulations and can provide proper care; or

(2) Placed with persons or facilities that can offer a level of care equal to or exceeding the standards and regulations, as determined by APHIS, even if the persons or facilities are not licensed by or registered with APHIS; or

95

(3) Euthanized.

(d) The dealer, exhibitor, intermediate handler, or carrier from whom the animals were confiscated must bear all costs incurred in performing the placement or euthanasia activities authorized by this section.

[54 FR 36147, Aug. 31, 1989, as amended at 66 FR 239, Jan. 3, 2001]

§ 2.130 - Minimum age requirements.

No dog or cat shall be delivered by any person to any carrier or intermediate handler for transportation, in commerce, or shall be transported in commerce by any person, except to a registered research facility, unless such dog or cat is at least eight (8) weeks of age and has been weaned.

§ 2.131 - Handling of animals.

(a) All licensees who maintain wild or exotic animals must demonstrate adequate experience and knowledge of the species they maintain.

(b)(1) Handling of all animals shall be done as expeditiously and carefully as possible in a manner that does not cause trauma, overheating, excessive cooling, behavioral stress, physical harm, or unnecessary discomfort.

(2)(i) Physical abuse shall not be used to train, work, or otherwise handle animals.

(ii) Deprivation of food or water shall not be used to train, work, or otherwise handle animals; *Provided, however,* That the short-term withholding of food or water from animals by exhibitors is allowed by these regulations as long as each of the animals affected receives its full dietary and nutrition requirements each day.

(c)(1) During public exhibition, any animal must be handled so there is minimal risk of harm to the animal and to the public, with sufficient distance and/or barriers between the animal and the general viewing public so as to assure the safety of animals and the public.

(2) Performing animals shall be allowed a rest period between performances at least equal to the time for one performance.

(3) Young or immature animals shall not be exposed to rough or excessive public handling or exhibited for periods of time which would be detrimental to their health or well-being.

(4) Drugs, such as tranquilizers, shall not be used to facilitate, allow, or provide for public handling of the animals.

(d)(1) Animals shall be exhibited only for periods of time and under conditions consistent with their good health and well-being.

(2) A responsible, knowledgeable, and readily identifiable employee or attendant must be present at all times during periods of public contact.

(3) During public exhibition, dangerous animals such as lions, tigers, wolves, bears, or elephants must be under the direct control and supervision of a knowledgeable and experienced animal handler.

(4) If public feeding of animals is allowed, the food must be provided by the animal facility and shall be appropriate to the type of animal and its nutritional needs and diet.

(e) When climatic conditions present a threat to an animal's health or well-being, appropriate measures must be taken to alleviate the impact of those conditions. An animal may never be subjected to any combination of temperature, humidity, and time that is detrimental to the animal's health or well-being, taking into consideration such factors as the animal's age, species, breed, overall health status, and acclimation.

[54 FR 36147, Aug. 31, 1989, as amended at 63 FR 10498, Mar. 4, 1998; 69 FR 42102, July 14, 2004]

§ 2.132 - Procurement of dogs, cats, and other animals; dealers.

(a) A class "B" dealer may obtain live random source dogs and cats only from:

(1) Other dealers who are licensed under the Act and in accordance with the regulations in part 2;

(2) State, county, or city owned and operated animal pounds or shelters; and

(3) A legal entity organized and operated under the laws of the State in which it is located as an animal pound or shelter, such as a humane shelter or contract pound.

(b) No person shall obtain live dogs, cats, or other animals by use of false pretenses, misrepresentation, or deception.

(c) Any dealer, exhibitor, research facility, carrier, or intermediate handler who also operates a private or contract animal pound or shelter shall comply with the following:

(1) The animal pound or shelter shall be located on premises that are physically separated from the licensed or registered facility. The animal housing facility of the pound or shelter shall not be adjacent to the licensed or registered facility.

(2) Accurate and complete records shall be separately maintained by the licensee or registrant and by the pound or shelter. The records shall be in accordance with §§ 2.75 and 2.76, unless the animals are lost or stray. If the animals are lost or stray, the pound or shelter records shall provide:

(i) An accurate description of the animal;

(ii) How, where, from whom, and when the dog or cat was obtained;

(iii) How long the dog or cat was held by the pound or shelter before being transferred to the dealer; and

(iv) The date the dog or cat was transferred to the dealer.

(3) Any dealer who obtains or acquires a live dog or cat from a private or contract pound or shelter, including a pound or shelter he or she operates, shall hold the dog or cat for a period of at least 10 full days, not including the day of acquisition, excluding time in transit, after acquiring the animal, and otherwise in accordance with § 2.101.

(d) No dealer or exhibitor shall knowingly obtain any dog, cat, or other animal from any person who is required to be licensed but who does not hold a current, valid, and unsuspended license. No dealer or exhibitor shall knowingly obtain any dog or cat from any person who is not licensed, other than a pound or shelter, without obtaining a certification that the animals were born and raised on that person's premises and, if the animals are for research purposes, that the person has sold fewer than 25 dogs and/or cats that year, or, if the animals are for use as pets, that the person does not maintain more than four breeding female dogs and/or cats.

(Approved by the Office of Management and Budget under control number 0579-0254)

[54 FR 36147, Aug. 31, 1989, as amended at 69 FR 42102, July 14, 2004; 80 FR 3463, Jan. 23, 2015]

§ 2.133 - Certification for random source dogs and cats.

(a) Each of the entities listed in paragraphs (a)(1) through (a)(3) of this section that acquire any live dog or cat shall, before selling or providing the live dog or cat to a dealer, hold and care for the dog or cat for a period of not less than 5 full days after acquiring the animal, not including the date of acquisition and excluding time in transit. This holding period shall include at least one Saturday. The provisions of this paragraph apply to:

(1) Each pound or shelter owned and operated by a State, county, or city;

(2) Each private pound or shelter established for the purpose of caring for animals, such as a humane society, or other organization that is under contract with a State, county, or city, that operates as a pound or shelter, and that releases animals on a voluntary basis; and

(3) Each research facility licensed by USDA as a dealer.

(b) A dealer shall not sell, provide, or make available to any person a live random source dog or cat unless the dealer provides the recipient of the dog or cat with certification that contains the following information:

(1) The name, address, USDA license number, and signature of the dealer;

(2) The name, address, USDA license or registration number, if such number exists, and signature of the recipient of the dog or cat;

(3) A description of each dog or cat being sold, provided, or made available that shall include:

PART 2
Subpart I

(i) The species and breed or type (for mixed breeds, estimate the two dominant breeds or types);

(ii) The sex;

(iii) The date of birth or, if unknown, then the approximate age;

(iv) The color and any distinctive markings; and

(v) The Official USDA-approved identification number of the animal. However, if the certification is attached to a certificate provided by a prior dealer which contains the required description, then only the official identification numbers are required;

(4) The name and address of the person, pound, or shelter from which the dog or cat was acquired by the dealer, and an assurance that the person, pound, or shelter was notified that the cat or dog might be used for research or educational purposes;

(5) The date the dealer acquired the dog or cat from the person, pound, or shelter referred to in paragraph (b)(4) of this section; and

(6) If the dealer acquired the dog or cat from a pound or shelter, a signed statement by the pound or shelter that it met the requirements of paragraph (a) of this section. This statement must at least describe the animals by their official USDA identification numbers. It may be incorporated within the certification if the dealer makes the certification at the time that the animals are acquired from the pound or shelter or it may be made separately and attached to the certification later. If made separately, it must include the same information describing each animal as is required in the certification. A photocopy of the statement will be regarded as a duplicate original.

(c) The original certification required under paragraph (b) of this section shall accompany the shipment of a live dog or cat to be sold, provided, or otherwise made available by the dealer.

(d) A dealer who acquires a live dog or cat from another dealer must obtain from that dealer the certification required by paragraph (b) of this section and must attach that certification (including any previously attached certification) to the certification which he or she provides pursuant to paragraph (b) of this section (a photocopy of the original certification will be deemed a duplicate original if the dealer does not dispose of all of the dogs or cats in a single transaction).

(e) A dealer who completes, provides, or receives a certification required under paragraph (b) of this section shall keep, maintain, and make available for APHIS inspection a copy of the certification for at least 1 year following disposition.

(f) A research facility which acquires any live random source dog or cat from a dealer must obtain the certification required under paragraph (b) of this section and shall keep, maintain, and make available for APHIS inspection the original for at least 3 years following disposition.

(g) In instances where a research facility transfers ownership of a live random source dog or cat acquired from a dealer to another research facility, a copy of the certification required by paragraph (b) of this section must accompany the dog or cat transferred. The research facility to which the dog or cat is transferred shall keep, maintain, and make available for APHIS inspection the copy of the certification for at least 3 years following disposition.

[58 FR 39129, July 22, 1993]

§ 2.134 - Contingency planning.

(a) Dealers, exhibitors, intermediate handlers, and carriers must develop, document, and follow an appropriate plan to provide for the humane handling, treatment, transportation, housing, and care of their animals in the event of an emergency or disaster (one which could reasonably be anticipated and expected to be detrimental to the good health and well-being of the animals in their possession). Such contingency plans must:

(1) Identify situations the licensee or registrant might experience that would trigger the need for the measures identified in a contingency plan to be put into action including, but not limited to, emergencies such as electrical outages, faulty HVAC systems, fires, mechanical breakdowns, and animal escapes, as well as natural disasters most likely to be experienced;

(2) Outline specific tasks required to be carried out in response to the identified emergencies or disasters including, but not limited to, detailed animal evacuation instructions or shelter-in-place instructions and provisions for providing backup sources of food and water as well as sanitation, ventilation, bedding, veterinary care, etc.;

(3) Identify a chain of command and who (by name or by position title) will be responsible for fulfilling these tasks; and

(4) Address how response and recovery will be handled in terms of materials, resources, and training needed.

(b) For current licensees and registrants, the contingency plan must be in place by July 29, 2013. For new dealers, exhibitors, intermediate handlers, and carriers licensed or registered after this date, the contingency plan must be in place prior to conducting regulated activities. The plan must be reviewed by the dealer, exhibitor, intermediate handler, or carrier on at least an annual basis to ensure that it adequately addresses the criteria listed in paragraph (a) of this section. Each licensee and registrant must maintain documentation of their annual reviews, including documenting any amendments or changes made to their plan since the previous year's review, such as changes made as a result of recently predicted, but historically unforeseen, circumstances (e.g., weather extremes). Contingency plans, as well as all annual review documentation and training records, must be made available to APHIS upon request. Traveling entities must carry a copy of their contingency plan with them at all times and make it available for APHIS inspection while in travel status. Dealers, exhibitors, intermediate handlers, and carriers maintaining or otherwise handling marine mammals in captivity must also comply with the requirements of § 3.101(b) of this subchapter.

(c) Dealers, exhibitors, intermediate handlers, and carriers must provide and document participation in and successful completion of training for personnel regarding their roles and responsibilities as outlined in the plan. For current licensees and registrants, training of dealer, exhibitor, intermediate handler, and carrier personnel must be completed by September 27, 2013. For new dealers, exhibitors, intermediate handlers, or carriers licensed or registered after July 29, 2013, training of personnel must be completed within 60 days of the dealer, exhibitor, intermediate handler, or carrier putting their contingency plan in place. Employees hired 30 days or more before their contingency plan is put in place must also be trained by that date. For employees hired less than 30 days before that date or after that date, training must be conducted within 30 days of their start date. Any changes to the plan as a result of the annual review must be communicated to employees through training which must be conducted within 30 days of making the changes.

[77 FR 76823, Dec. 31, 2012]

Effective Date Note: *At 78 FR 46255, July 31, 2013, §2.134 was stayed indefinitely, effective July 31, 2013.*

Subpart J—Importation of Live Dogs

§ 2.150 Import Permit

(a) No person shall import a live dog from any part of the world into the continental United States or Hawaii for purposes of resale, research or veterinary treatment unless the dog is accompanied by an import permit issued by APHIS and is imported into the continental United States or Hawaii within 30 days after the proposed date of arrival stated in the import permit.

(b) An application for an import permit must be submitted to the Animal and Plant Health Inspection Service, Animal Care, 4700 River Road Unit 84, Riverdale, MD 20737–1234 or though Animal Care's Web site (*http://www. aphis.usda.gov/animal_welfare/*). Application forms for import permits may be obtained from Animal Care at the address listed above.

(c) The completed application must include the following information:

(1) The name and address of the person intending to export the dog(s) to the continental United States or Hawaii;

(2) The name and address of the person intending to import the dog(s) into the continental United States or Hawaii;

(3) The number of dogs to be imported and the breed, sex, age, color, markings, and other identifying information of each dog;

(4) The purpose of the importation;

(5) The port of embarkation and the mode of transportation;

(6) The port of entry in the United States;

(7) The proposed date of arrival in the continental United States or Hawaii; and

(8) The name and address of the person to whom the dog(s) will be delivered in the continental United States or Hawaii and, if the dog(s) is or are imported for research purposes, the USDA registration number of the research facility where the dog will be used for research, tests, or experiments.

(d) After receipt and review of the application by APHIS, an import permit indicating the applicable conditions for importation under this subpart may be issued for the importation of the dog(s) described in the application if such dog(s) appears to be eligible to be imported. Even though an import permit has been issued for the importation of a dog, the dog may only be imported if all applicable requirements of this subpart and any other applicable regulations of this subchapter and any other statute or regulation of any State or of the United States are met.

(Approved by the Office of Management and Budget under control number 0579–0379)

§ 2.151 Certifications.

(a) *Required certificates.* Except as provided in paragraph (b) of this section, no person shall import a live dog from any part of the world into the continental United States or Hawaii for purposes of resale, research, or veterinary treatment unless the following conditions are met:

(1) *Health certificate.* Each dog is accompanied by an original health certificate issued in English by a licensed veterinarian with a valid license to practice veterinary medicine in the country of export that:

(i) Specifies the name and address of the person intending to import the dog into the continental United States or Hawaii;

(ii) Identifies the dog on the basis of breed, sex, age, color, markings, and other identifying information;

(iii) States that the dog is at least 6 months of age;

(iv) States that the dog was vaccinated, not more than 12 months before the date of arrival at the U.S. port, for distemper, hepatitis, leptospirosis, parvovirus, and parainfluenza virus (DHLPP) at a frequency that provides continuous protection of the dog from those diseases and is in accordance with currently accepted practices as cited in veterinary medicine reference guides;

(v) States that the dog is in good health (i.e., free of any infectious disease or physical abnormality which would endanger the dog or other animals or endanger public health, including, but not limited to, parasitic infection, emaciation, lesions of the skin, nervous system disturbances, jaundice, or diarrhea); and

(vi) Bears the signature and the license number of the veterinarian issuing the certificate.

(2) *Rabies vaccination certificate.* Each dog is accompanied by a valid rabies vaccination certificate that was issued in English by a licensed veterinarian with a valid license to practice veterinary medicine in the country of export for the dog not less than 3 months of age at the time of vaccination that:[6]

(i) Specifies the name and address of the person intending to import the dog into the continental United States or Hawaii;

(ii) Identifies the dog on the basis of breed, sex, age, color markings and other identifying information;

(iii) Specifies a date of rabies vaccination at least 30 days before the date of arrival of the dog at a U.S. port;

(iv) Specifies a date of expiration of the vaccination which is after the date of arrival of the dog at a U.S. port. If no date of expiration is

---

6 *Alternatively, this requirement can be met by providing an exact copy of the rabies vaccination certificate if so required under the Public Health Service regulations in 42 CFR 71.51.*

specified, then the date of vaccination shall be no more than 12 months before the date of arrival at a U.S. port; and

(v) Bears the signature and the license number of the veterinarian issuing the certificate.

(b) *Exceptions.*

(1) *Research.* The provisions of paragraphs (a)(1)(iii), (a)(1)(iv), (a)(1)(v), and/or (a)(2) of this section do not apply to any person who imports a live dog from any part of the world into the continental United States or Hawaii for use in research, tests, or experiments at a research facility, provided that: Such person submits satisfactory evidence to Animal Care at the time of his or her application for an import permit that the specific provision(s) would interfere with the dog's use in such research, tests, or experiments in accordance with a research proposal and the proposal has been approved by the research facility IACUC.

PART 2
Subpart J

(2) *Veterinary care.* The provisions of paragraphs (a)(1)(iii) through (a)(1)(v) and (a)(2) of this section do not apply to any person who imports a live dog from any part of the world into the continental United States or Hawaii for veterinary treatment by a licensed veterinarian, provided that:

(i) The original health certificate required in paragraph (a)(1) of this section states that the dog is in need of veterinary treatment that cannot be obtained in the country of export and states the name and address of the licensed veterinarian in the United States who intends to provide the dog such veterinary treatment; and

(ii) The person who imports the dog completes a veterinary treatment agreement with Animal Care at the time of application for an import permit and confines the animal until the conditions specified in the agreement are met. Such conditions may include determinations by the licensed veterinarian in the United States that the dog is in good health, has been adequately vaccinated against DHLPP and rabies, and is at least six months of age. The person importing the dog shall bear the expense of veterinary treatment and confinement.

(3) *Dogs imported into Hawaii from the British Isles, Australia, Guam, or New Zealand.* The provisions of paragraph (a)(1)(iii) of this section do not apply to any person who lawfully imports a live dog into the State of Hawaii from the British Isles, Australia, Guam, or New Zealand in compliance with the applicable regulations of the State of Hawaii, provided that the dog is not transported out of the State of Hawaii for purposes of resale at less than 6 months of age.

(Approved by the Office of Management and Budget under control number 0579–0379)

105

§ 2.152 Notification of arrival.

Upon the arrival of a dog at the port of first arrival in the continental United States or Hawaii, the person intending to import the dog, or his or her agent, must present the import permit and any applicable certifications and veterinary treatment agreement required by this subpart to the collector of customs for use at that port.

PART 2
Subpart J

(Approved by the Office of Management and Budget under control number 0579–0379)

§ 2.153 Dogs refused entry.

Any dog refused entry into the continental United States or Hawaii for noncompliance with the requirements of this subpart may be removed from the continental United States or Hawaii or may be seized and the person intending to import the dog shall provide for the care (including appropriate veterinary care), forfeiture, and adoption of the dog, at his or her expense.

PART 3 – STANDARDS

Subpart A – Specifications for the Humane Handling, Care, Treatment, and Transportation of Dogs and Cats

FACILITIES AND OPERATING STANDARDS
§ 3.1 Housing facilities, general.
§ 3.2 Indoor housing facilities.
§ 3.3 Sheltered housing facilities.
§ 3.4 Outdoor housing facilities.
§ 3.5 Mobile or traveling housing facilities.
§ 3.6 Primary enclosures.

ANIMAL HEALTH AND HUSBANDRY STANDARDS
§ 3.7 Compatible grouping.
§ 3.8 Exercise for dogs.
§ 3.9 Feeding.
§ 3.10 Watering.
§ 3.11 Cleaning, sanitization, housekeeping, and pest control.
§ 3.12 Employees.

TRANSPORTATION STANDARDS
§ 3.13 Consignments to carriers and intermediate handlers.
§ 3.14 Primary enclosures used to transport live dogs and cats.
§ 3.15 Primary conveyances (motor vehicle, rail, air, and marine).
§ 3.16 Food and water requirements.
§ 3.17 Care in transit.
§ 3.18 Terminal facilities.
§ 3.19 Handling.

Subpart B – Specifications for the Humane Handling, Care, Treatment, and Transportation of Guinea Pigs and Hamsters

FACILITIES AND OPERATING STANDARDS
§ 3.25 Facilities, general.
§ 3.26 Facilities, indoor.
§ 3.27 Facilities, outdoor.
§ 3.28 Primary enclosures.

ANIMAL HEALTH AND HUSBANDRY STANDARDS
§ 3.29 Feeding.
§ 3.30 Watering.
§ 3.31 Sanitation.

Animal Welfare Regulations, USDA

FACILITIES AND OPERATING STANDARDS

ANIMAL HEALTH AND HUSBANDRY STANDARDS

TRANSPORTATION STANDARDS

Subpart E – Specifications for the Humane Handling, Care, Treatment, and Transportation of Marine Mammals

FACILITIES AND OPERATING STANDARDS

ANIMAL HEALTH AND HUSBANDRY STANDARDS

TRANSPORTATION STANDARDS

Subpart F – Specifications for the Humane Handling, Care, Treatment, and Transportation of Warm-blooded Animals Other Than Dogs, Cats, Rabbits, Hamsters, Guinea Pigs, Nonhuman Primates, and Marine Mammals

PART 3
Table of Contents

FACILITIES AND OPERATING STANDARDS

ANIMAL HEALTH AND HUSBANDRY STANDARDS

TRANSPORTATION STANDARDS

Authority: 7 U.S.C. 2131-2159; 7 CFR 2.22, 2.80, and 371.7.

Source: 32 FR 3273, Feb. 24, 1967, unless otherwise noted.

Subpart A – Specifications for the Humane Handling, Care, Treatment, and Transportation of Dogs and Cats[1]

Source: 56 FR 6486, Feb. 15, 1991, unless otherwise noted.

FACILITIES AND OPERATING STANDARDS

§ 3.1 - Housing facilities, general.

(a) *Structure; construction.* Housing facilities for dogs and cats must be designed and constructed so that they are structurally sound. They must be kept in good repair, and they must protect the animals from injury, contain the animals securely, and restrict other animals from entering.

(b) *Condition and site.* Housing facilities and areas used for storing animal food or bedding must be free of any accumulation of trash, waste material, junk, weeds, and other discarded materials. Animal areas inside of housing facilities must be kept neat and free of clutter, including equipment, furniture, and stored material, but may contain materials actually used and necessary for cleaning the area, and fixtures or equipment necessary for proper husbandry practices and research needs. Housing facilities other than those maintained by research facilities and Federal research facilities must be physically separated from any other business. If a housing facility is located on the same premises as another business, it must be physically separated from the other business so that animals the size of dogs, skunks, and raccoons are prevented from entering it.

(c) *Surfaces.*

(1) **General requirements.** The surfaces of housing facilities – including houses, dens, and other furniture-type fixtures and objects within the facility – must be constructed in a manner and made of materials that allow them to be readily cleaned and sanitized, or removed or replaced when worn or soiled. Interior surfaces and any surfaces that come in contact with dogs or cats must:

(i) Be free of excessive rust that prevents the required cleaning and sanitization, or that affects the structural strength of the surface; and

(ii) Be free of jagged edges or sharp points that might injure the animals.

(2) *Maintenance and replacement of surfaces.* All surfaces must be maintained on a regular basis. Surfaces of housing facilities – including houses, dens, and other furniture-type fixtures and objects within the facility

PART 3
Subpart A

---

[1] *These minimum standards apply only to live dogs and cats, unless stated otherwise.*

111

– that cannot be readily cleaned and sanitized, must be replaced when worn or soiled.

(3) *Cleaning.* Hard surfaces with which the dogs or cats come in contact must be spot-cleaned daily and sanitized in accordance with § 3.11(b) of this subpart to prevent accumulation of excreta and reduce disease hazards. Floors made of dirt, absorbent bedding, sand, gravel, grass, or other similar material must be raked or spot-cleaned with sufficient frequency to ensure all animals the freedom to avoid contact with excreta. Contaminated material must be replaced whenever this raking and spot-cleaning is not sufficient to prevent or eliminate odors, insects, pests, or vermin infestation. All other surfaces of housing facilities must be cleaned and sanitized when necessary to satisfy generally accepted husbandry standards and practices. Sanitization may be done using any of the methods provided in § 3.11(b)(3) for primary enclosures.

(d) *Water and electric power.* The housing facility must have reliable electric power adequate for heating, cooling, ventilation, and lighting, and for carrying out other husbandry requirements in accordance with the regulations in this subpart. The housing facility must provide adequate running potable water for the dogs' and cats' drinking needs, for cleaning, and for carrying out other husbandry requirements.

(e) *Storage.* Supplies of food and bedding must be stored in a manner that protects the supplies from spoilage, contamination, and vermin infestation. The supplies must be stored off the floor and away from the walls, to allow cleaning underneath and around the supplies. Foods requiring refrigeration must be stored accordingly, and all food must be stored in a manner that prevents contamination and deterioration of its nutritive value. All open supplies of food and bedding must be kept in leakproof containers with tightly fitting lids to prevent contamination and spoilage. Only food and bedding that is currently being used may be kept in the animal areas. Substances that are toxic to the dogs or cats but are required for normal husbandry practices must not be stored in food storage and preparation areas, but may be stored in cabinets in the animal areas.

(f) *Drainage and waste disposal.* Housing facility operators must provide for regular and frequent collection, removal, and disposal of animal and food wastes, bedding, debris, garbage, water, other fluids and wastes, and dead animals, in a manner that minimizes contamination and disease risks. Housing facilities must be equipped with disposal facilities and drainage systems that are constructed and operated so that animal waste and water are rapidly eliminated and animals stay dry. Disposal and drainage systems must minimize vermin and pest infestation, insects, odors, and disease hazards. All drains must be properly constructed, installed, and maintained. If closed drainage systems are used, they must be equipped with traps and prevent the backflow of gases and the backup of sewage onto the floor. If the facility uses sump or settlement ponds, or other similar systems for drainage and animal waste disposal, the system must be located far enough away from the animal area of the housing facility to prevent odors, diseases, pests, and vermin infestation. Standing puddles of water in animal enclosures must be drained or mopped up so that the animals stay dry. Trash containers in housing facilities and in food storage and food preparation areas must be leak-proof and must have tightly fitted lids on them at all times. Dead animals, animal parts, and animal waste must not be kept in food storage or food preparation areas, food freezers, food refrigerators, or animal areas.

(g) *Washrooms and sinks.* Washing facilities such as washrooms, basins, sinks, or showers must be provided for animal caretakers and must be readily accessible.

§ 3.2 - Indoor housing facilities.

(a) *Heating, cooling, and temperature.* Indoor housing facilities for dogs and cats must be sufficiently heated and cooled when necessary to protect the dogs and cats from temperature or humidity extremes and to provide for their health and well-being. When dogs or cats are present, the ambient temperature in the facility must not fall below 50 °F (10 °C) for dogs and cats not acclimated to lower temperatures, for those breeds that cannot tolerate lower temperatures without stress or discomfort (such as short-haired breeds), and for sick, aged, young, or infirm dogs and cats, except as approved by the attending veterinarian. Dry bedding, solid resting boards, or other methods of conserving body heat must be provided when temperatures are below 50 °F (10 °C). The ambient temperature must not fall below 45 °F (7.2 °C) for more than 4 consecutive hours when dogs or cats are present, and must not rise above 85 °F (29.5 °C) for more than 4 consecutive hours when dogs or cats are present. The preceding requirements are in addition to, not in place of, all other requirements pertaining to climatic conditions in parts 2 and 3 of this chapter.

PART 3
Subpart A

(b) *Ventilation.* Indoor housing facilities for dogs and cats must be sufficiently ventilated at all times when dogs or cats are present to provide for their health and well-being, and to minimize odors, drafts, ammonia levels, and moisture condensation. Ventilation must be provided by windows, vents, fans, or air conditioning. Auxiliary ventilation, such as fans, blowers, or air conditioning must be provided when the ambient temperature is 85 °F (29.5 °C) or higher. The relative humidity must be maintained at a level that ensures the health and well-being of the dogs or cats housed therein, in accordance with the directions of the attending veterinarian and generally accepted professional and husbandry practices.

(c) *Lighting.* Indoor housing facilities for dogs and cats must be lighted well enough to permit routine inspection and cleaning of the facility, and observation of the dogs and cats. Animal areas must be provided a regular diurnal lighting cycle of either natural or artificial light. Lighting must be uniformly diffused throughout animal facilities and provide sufficient illumination to aid in maintaining good housekeeping practices, adequate cleaning, adequate inspection of animals, and for the well-being of the animals. Primary enclosures must be placed so as to protect the dogs and cats from excessive light.

(d) *Interior surfaces.* The floors and walls of indoor housing facilities, and any other surfaces in contact with the animals, must be impervious to moisture. The ceilings of indoor housing facilities must be impervious to moisture or be replaceable (e.g., a suspended ceiling with replaceable panels).

[56 FR 6486, Feb. 15, 1991, as amended at 63 FR 10498, Mar. 4, 1998]

PART 3
Subpart A

§ 3.3 - Sheltered housing facilities.

(a) *Heating, cooling, and temperature.* The sheltered part of sheltered housing facilities for dogs and cats must be sufficiently heated and cooled when necessary to protect the dogs and cats from temperature or humidity extremes and to provide for their health and well-being. The ambient temperature in the sheltered part of the facility must not fall below 50 °F (10 °C) for dogs and cats not acclimated to lower temperatures, for those breeds that cannot tolerate lower temperatures without stress and discomfort (such as short-haired breeds), and for sick, aged, young, or infirm dogs or cats, except as approved by the attending veterinarian. Dry bedding, solid resting boards, or other methods of conserving body heat must be provided when temperatures are below 50 °F (10 °C). The ambient temperature must not fall below 45 °F (7.2 °C) for more than 4 consecutive hours when dogs or cats are present, and must not rise above 85 °F (29.5 °C) for more than 4 consecutive hours when dogs or cats are present. The preceding requirements are in addition to, not in place of, all other requirements pertaining to climatic conditions in parts 2 and 3 of this chapter.

(b) *Ventilation.* The enclosed or sheltered part of sheltered housing facilities for dogs and cats must be sufficiently ventilated when dogs or cats are present to provide for their health and well-being, and to minimize odors, drafts, ammonia levels, and moisture condensation. Ventilation must be provided by windows, doors, vents, fans, or air conditioning. Auxiliary ventilation, such as fans, blowers, or air-conditioning, must be provided when the ambient temperature is 85 °F (29.5 °C) or higher.

(c) *Lighting.* Sheltered housing facilities for dogs and cats must be lighted well enough to permit routine inspection and cleaning of the facility, and

114

observation of the dogs and cats. Animal areas must be provided a regular diurnal lighting cycle of either natural or artificial light. Lighting must be uniformly diffused throughout animal facilities and provide sufficient illumination to aid in maintaining good housekeeping practices, adequate cleaning, adequate inspection of animals, and for the well-being of the animals. Primary enclosures must be placed so as to protect the dogs and cats from excessive light.

(d) *Shelter from the elements.* Dogs and cats must be provided with adequate shelter from the elements at all times to protect their health and well-being. The shelter structures must be large enough to allow each animal to sit, stand, and lie in a normal manner and to turn about freely.

(e) *Surfaces.*

(1) The following areas in sheltered housing facilities must be impervious to moisture:

(i) Indoor floor areas in contact with the animals;

(ii) Outdoor floor areas in contact with the animals, when the floor areas are not exposed to the direct sun, or are made of a hard material such as wire, wood, metal, or concrete; and

PART 3
Subpart A

(iii) All walls, boxes, houses, dens, and other surfaces in contact with the animals.

(2) Outside floor areas in contact with the animals and exposed to the direct sun may consist of compacted earth, absorbent bedding, sand, gravel, or grass.

[56 FR 6486, Feb. 15, 1991, as amended at 63 FR 10498, Mar. 4, 1998]

§ 3.4 - Outdoor housing facilities.

(a) *Restrictions.*

(1) The following categories of dogs or cats must not be kept in outdoor facilities, unless that practice is specifically approved by the attending veterinarian:

(i) Dogs or cats that are not acclimated to the temperatures prevalent in the area or region where they are maintained;

(ii) Breeds of dogs or cats that cannot tolerate the prevalent temperatures of the area without stress or discomfort (such as short-haired breeds in cold climates); and

(iii) Sick, infirm, aged or young dogs or cats.

(2) When their acclimation status is unknown, dogs and cats must not be kept in outdoor facilities when the ambient temperature is less than 50 °F (10 °C).

(b) *Shelter from the elements.* Outdoor facilities for dogs or cats must include one or more shelter structures that are accessible to each animal in each outdoor facility, and that are large enough to allow each animal in the shelter structure to sit, stand, and lie in a normal manner, and to turn about freely. In addition to the shelter structures, one or more separate outside areas of shade must be provided, large enough to contain all the animals at one time and protect them from the direct rays of the sun. Shelters in outdoor facilities for dogs or cats must contain a roof, four sides, and a floor, and must:

(1) Provide the dogs and cats with adequate protection and shelter from the cold and heat;

(2) Provide the dogs and cats with protection from the direct rays of the sun and the direct effect of wind, rain, or snow;

(3) Be provided with a wind break and rain break at the entrance; and

(4) Contain clean, dry, bedding material if the ambient temperature is below 50 °F (10 °C). Additional clean, dry bedding is required when the temperature is 35 °F (1.7 °C) or lower.

(c) *Construction.* Building surfaces in contact with animals in outdoor housing facilities must be impervious to moisture. Metal barrels, cars, refrigerators or freezers, and the like must not be used as shelter structures. The floors of outdoor housing facilities may be of compacted earth, absorbent bedding, sand, gravel, or grass, and must be replaced if there are any prevalent odors, diseases, insects, pests, or vermin. All surfaces must be maintained on a regular basis. Surfaces of outdoor housing facilities – including houses, dens, etc. – that cannot be readily cleaned and sanitized must be replaced when worn or soiled.

PART 3
Subpart A

§ 3.5 - Mobile or traveling housing facilities.

(a) *Heating, cooling, and temperature.* Mobile or traveling housing facilities for dogs and cats must be sufficiently heated and cooled when necessary to protect the dogs and cats from temperature or humidity extremes and to provide for their health and well-being. The ambient temperature in the mobile or traveling housing facility must not fall below 50 °F (10 °C) for dogs and cats not acclimated to lower temperatures, for those breeds that cannot tolerate lower temperatures without stress or discomfort (such as short-haired breeds), and for sick, aged, young, or infirm dogs and cats. Dry bedding, solid resting boards, or other methods of conserving body heat must be provided when temperatures are below 50 °F (10 °C). The ambient temperature must not fall below 45 °F (7.2 °C) for more than 4 consecutive hours when dogs or cats are present, and must not exceed 85 °F (29.5 °C) for more than 4 consecutive hours when dogs or cats are present. The preceding requirements are in addition to, not in place of, all other requirements pertaining to climatic conditions in parts 2 and 3 of this chapter.

(b) *Ventilation.* Mobile or traveling housing facilities for dogs and cats must be sufficiently ventilated at all times when dogs or cats are present to provide for the health and well-being of the animals, and to minimize odors, drafts, ammonia levels, moisture condensation, and exhaust fumes. Ventilation must be provided by means of windows, doors, vents, fans, or air conditioning. Auxiliary ventilation, such as fans, blowers, or air conditioning, must be provided when the ambient temperature within the animal housing area is 85 °F (29.5 °C) or higher.

(c) *Lighting.* Mobile or traveling housing facilities for dogs and cats must be lighted well enough to permit proper cleaning and inspection of the facility, and observation of the dogs and cats. Animal areas must be provided a regular diurnal lighting cycle of either natural or artificial light. Lighting must be uniformly diffused throughout animal facilities and provide sufficient illumination to aid in maintaining good housekeeping practices, adequate cleaning, adequate inspection of animals, and for the well-being of the animals.

[32 FR 3273, Feb. 24, 1967, as amended at 63 FR 10498, Mar. 4, 1998]

PART 3
Subpart A

§ 3.6 - Primary enclosures.

Primary enclosures for dogs and cats must meet the following minimum requirements:

(a) *General requirements.*

(1) Primary enclosures must be designed and constructed of suitable materials so that they are structurally sound. The primary enclosures must be kept in good repair.

(2) Primary enclosures must be constructed and maintained so that they:

(i) Have no sharp points or edges that could injure the dogs and cats;

(ii) Protect the dogs and cats from injury;

(iii) Contain the dogs and cats securely;

(iv) Keep other animals from entering the enclosure;

(v) Enable the dogs and cats to remain dry and clean;

(vi) Provide shelter and protection from extreme temperatures and weather conditions that may be uncomfortable or hazardous to all the dogs and cats;

(vii) Provide sufficient shade to shelter all the dogs and cats housed in the primary enclosure at one time;

(viii) Provide all the dogs and cats with easy and convenient access to clean food and water;

(ix) Enable all surfaces in contact with the dogs and cats to be readily cleaned and sanitized in accordance with § 3.11(b) of this subpart, or be replaceable when worn or soiled;

(x) Have floors that are constructed in a manner that protects the dogs' and cats' feet and legs from injury, and that, if of mesh or slatted construction, do not allow the dogs' and cats' feet to pass through any openings in the floor;

(xi) Provide sufficient space to allow each dog and cat to turn about freely, to stand, sit, and lie in a comfortable, normal position, and to walk in a normal manner; and

(xii) Primary enclosures constructed on or after February 20, 1998 and floors replaced on or after that date, must comply with the requirements in this paragraph (a)(2). On or after January 21, 2000, all primary enclosures must be in compliance with the requirements in this paragraph (a)(2). If the suspended floor of a primary enclosure is constructed of metal strands, the strands must either be greater than $\frac{1}{8}$ of an inch in diameter (9 gauge) or coated with a material such as plastic or fiberglass. The suspended floor of any primary enclosure must be strong enough so that the floor does not sag or bend between the structural supports.

(b) *Additional requirements for cats.*

(1) *Space.* Each cat, including weaned kittens, that is housed in any primary enclosure must be provided minimum vertical space and floor space as follows:

(i) Prior to February 15, 1994 each cat housed in any primary enclosure shall be provided a minimum of 2½ square feet of floor space;

(ii) On and after February 15, 1994:

(A) Each primary enclosure housing cats must be at least 24 in. high (60.96 cm);

(B) Cats up to and including 8.8 lbs (4 kg) must be provided with at least 3.0 ft² (0.28 m²);

(C) Cats over 8.8 lbs (4 kg) must be provided with at least 4.0 ft² (0.37 m²);

(iii) Each queen with nursing kittens must be provided with an additional amount of floor space, based on her breed and behavioral characteristics, and in accordance with generally accepted husbandry practices. If the additional amount of floor space for each nursing kitten is equivalent to less than 5 percent of the minimum requirement for the queen, such housing must be approved by the attending veterinarian in the case of a research facility, and, in the case of dealers and exhibitors, such housing must be approved by the Administrator; and

(iv) The minimum floor space required by this section is exclusive of any food or water pans. The litter pan may be considered part of the floor space if properly cleaned and sanitized.

(2) *Compatibility.* All cats housed in the same primary enclosure must be compatible, as determined by observation. Not more than 12 adult non-conditioned cats may be housed in the same primary enclosure. Queens in heat may not be housed in the same primary enclosure with sexually mature

males, except for breeding. Except when maintained in breeding colonies, queens with litters may not be housed in the same primary enclosure with other adult cats, and kittens under 4 months of age may not be housed in the same primary enclosure with adult cats, other than the dam or foster dam. Cats with a vicious or aggressive disposition must be housed separately.

(3) *Litter.* In all primary enclosures, a receptacle containing sufficient clean litter must be provided to contain excreta and body wastes.

(4) *Resting surfaces.* Each primary enclosure housing cats must contain a resting surface or surfaces that, in the aggregate, are large enough to hold all the occupants of the primary enclosure at the same time comfortably. The resting surfaces must be elevated, impervious to moisture, and be able to be easily cleaned and sanitized, or easily replaced when soiled or worn. Low resting surfaces that do not allow the space under them to be comfortably occupied by the animal will be counted as part of the floor space.

(5) *Cats in mobile or traveling shows or acts.* Cats that are part of a mobile or traveling show or act may be kept, while the show or act is traveling from one temporary location to another, in transport containers that comply with all requirements of § 3.14 of this subpart other than the marking requirements in § 3.14(a)(6) of this subpart. When the show or act is not traveling, the cats must be placed in primary enclosures that meet the minimum requirements of this section.

PART 3
Subpart A

(c) *Additional requirements for dogs.*

(1) *Space.*

(i) Each dog housed in a primary enclosure (including weaned puppies) must be provided a minimum amount of floor space, calculated as follows: Find the mathematical square of the sum of the length of the dog in inches (measured from the tip of its nose to the base of its tail) plus 6 inches; then divide the product by 144. The calculation is: (length of dog in inches + 6) × (length of dog in inches + 6) = required floor space in square inches. Required floor space in inches/144 = required floor space in square feet.

(ii) Each bitch with nursing puppies must be provided with an additional amount of floor space, based on her breed and behavioral characteristics, and in accordance with generally accepted husbandry practices as determined by the attending veterinarian. If the additional amount of floor space for each nursing puppy is less than 5 percent of the minimum requirement for the bitch, such housing must be approved by the attending veterinarian in the case of a research facility, and, in the case of dealers and exhibitors, such housing must be approved by the Administrator.

(iii) The interior height of a primary enclosure must be at least 6 inches higher than the head of the tallest dog in the enclosure when it is in a normal standing position: *Provided* That, prior to February 15, 1994, each dog must be able to stand in a comfortable normal position.

(2) *Compatibility.* All dogs housed in the same primary enclosure must be compatible, as determined by observation. Not more than 12 adult non-conditioned dogs may be housed in the same primary enclosure. Bitches in heat may not be housed in the same primary enclosure with sexually mature males, except for breeding. Except when maintained in breeding colonies, bitches with litters may not be housed in the same primary enclosure with other adult dogs, and puppies under 4 months of age may not be housed in the same primary enclosure with adult dogs, other than the dam or foster dam. Dogs with a vicious or aggressive disposition must be housed separately.

(3) *Dogs in mobile or traveling shows or acts.* Dogs that are part of a mobile or traveling show or act may be kept, while the show or act is traveling from one temporary location to another, in transport containers that comply with all requirements of § 3.14 of this subpart other than the marking requirements in § 3.14(a)(6) of this subpart. When the show or act is not traveling, the dogs must be placed in primary enclosures that meet the minimum requirements of this section.

(4) *Prohibited means of primary enclosure.* Permanent tethering of dogs is prohibited for use as primary enclosure. Temporary tethering of dogs is prohibited for use as primary enclosure unless approval is obtained from APHIS.

(d) Innovative primary enclosures not precisely meeting the floor area and height requirements provided in paragraphs (b)(1) and (c)(1) of this section, but that provide the dogs or cats with a sufficient volume of space and the opportunity to express species-typical behavior, may be used at research facilities when approved by the Committee, and by dealers and exhibitors when approved by the Administrator.

(Approved by the Office of Management and Budget under control number 0579-0093)

[56 FR 6486, Feb. 15, 1991, as amended at 62 FR 43275, Aug. 13, 1997; 63 FR 3023, Jan. 21, 1998; 63 FR 37482, July 13, 1998]

ANIMAL HEALTH AND HUSBANDRY STANDARDS

§ 3.7 - Compatible grouping.

Dogs and cats that are housed in the same primary enclosure must be compatible, with the following restrictions:

(a) Females in heat (estrus) may not be housed in the same primary enclosure with males, except for breeding purposes;

(b) Any dog or cat exhibiting a vicious or overly aggressive disposition must be housed separately;

(c) Puppies or kittens 4 months of age or less may not be housed in the same primary enclosure with adult dogs or cats other than their dams or foster dams, except when permanently maintained in breeding colonies;

(d) Dogs or cats may not be housed in the same primary enclosure with any other species of animals, unless they are compatible; and

(e) Dogs and cats that have or are suspected of having a contagious disease must be isolated from healthy animals in the colony, as directed by the attending veterinarian. When an entire group or room of dogs and cats is known to have or believed to be exposed to an infectious agent, the group may be kept intact during the process of diagnosis, treatment, and control.

§ 3.8 - Exercise for dogs.

Dealers, exhibitors, and research facilities must develop, document, and follow an appropriate plan to provide dogs with the opportunity for exercise. In addition, the plan must be approved by the attending veterinarian. The plan must include written standard procedures to be followed in providing the opportunity for exercise. The plan must be made available to APHIS upon request, and, in the case of research facilities, to officials of any pertinent funding Federal agency. The plan, at a minimum, must comply with each of the following:

PART 3
Subpart A

(a) *Dogs housed individually.* Dogs over 12 weeks of age, except bitches with litters, housed, held, or maintained by any dealer, exhibitor, or research facility, including Federal research facilities, must be provided the opportunity for exercise regularly if they are kept individually in cages, pens, or runs that provide less than two times the required floor space for that dog, as indicated by § 3.6(c)(1) of this subpart.

(b) *Dogs housed in groups.* Dogs over 12 weeks of age housed, held, or maintained in groups by any dealer, exhibitor, or research facility, including Federal research facilities, do not require additional opportunity for exercise regularly if they are maintained in cages, pens, or runs that provide in total at least 100 percent of the required space for each dog if maintained separately. Such animals may be maintained in compatible groups, unless:

(1) Housing in compatible groups is not in accordance with a research proposal and the proposal has been approved by the research facility Committee;

(2) In the opinion of the attending veterinarian, such housing would adversely affect the health or well-being of the dog(s); or

(3) Any dog exhibits aggressive or vicious behavior.

(c) *Methods and period of providing exercise opportunity.*

(1) The frequency, method, and duration of the opportunity for exercise shall be determined by the attending veterinarian and, at research facilities, in consultation with and approval by the Committee.

(2) Dealers, exhibitors, and research facilities, in developing their plan, should consider providing positive physical contact with humans that encourages exercise through play or other similar activities. If a dog is housed, held, or maintained at a facility without sensory contact with another dog, it must be provided with positive physical contact with humans at least daily.

(3) The opportunity for exercise may be provided in a number of ways, such as:

(i) Group housing in cages, pens or runs that provide at least 100 percent of the required space for each dog if maintained separately under the minimum floor space requirements of § 3.6(c)(1) of this subpart;

PART 3
Subpart A

(ii) Maintaining individually housed dogs in cages, pens, or runs that provide at least twice the minimum floor space required by § 3.6(c)(1) of this subpart;

(iii) Providing access to a run or open area at the frequency and duration prescribed by the attending veterinarian; or

(iv) Other similar activities.

(4) Forced exercise methods or devices such as swimming, treadmills, or carousel-type devices are unacceptable for meeting the exercise requirements of this section.

(d) *Exemptions.*

(1) If, in the opinion of the attending veterinarian, it is inappropriate for certain dogs to exercise because of their health, condition, or well-being, the dealer, exhibitor, or research facility may be exempted from meeting the requirements of this section for those dogs. Such exemption must be documented by the attending veterinarian and, unless the basis for exemption is a permanent condition, must be reviewed at least every 30 days by the attending veterinarian.

(2) A research facility may be exempted from the requirements of this section if the principal investigator determines for scientific reasons set forth in the research proposal that it is inappropriate for certain dogs to exercise. Such exemption must be documented in the Committee-approved proposal and must be reviewed at appropriate intervals as determined by the Committee, but not less than annually.

(3) Records of any exemptions must be maintained and made available to USDA officials or any pertinent funding Federal agency upon request.

(Approved by the Office of Management and Budget under control number 0579-0093)

§ 3.9 - Feeding.

(a) Dogs and cats must be fed at least once each day, except as otherwise might be required to provide adequate veterinary care. The food must be uncontaminated, wholesome, palatable, and of sufficient quantity and nutritive value to maintain the normal condition and weight of the animal. The diet must be appropriate for the individual animal's age and condition.

(b) Food receptacles must be used for dogs and cats, must be readily accessible to all dogs and cats, and must be located so as to minimize contamination by excreta and pests, and be protected from rain and snow. Feeding pans must either be made of a durable material that can be easily cleaned and sanitized or be disposable. If the food receptacles are not disposable, they must be kept clean and must be sanitized in accordance with § 3.11(b) of this subpart. Sanitization is achieved by using one of the methods described in § 3.11(b)(3) of this subpart. If the food receptacles are disposable, they must be discarded after one use. Self-feeders may be used for the feeding of dry food. If self-feeders are used, they must be kept clean and must be sanitized in accordance with § 3.11(b) of this subpart. Measures must be taken to ensure that there is no molding, deterioration, and caking of feed.

PART 3
Subpart A

§ 3.10 - Watering.

If potable water is not continually available to the dogs and cats, it must be offered to the dogs and cats as often as necessary to ensure their health and well-being, but not less than twice daily for at least 1 hour each time, unless restricted by the attending veterinarian. Water receptacles must be kept clean and sanitized in accordance with § 3.11(b) of this subpart, and before being used to water a different dog or cat or social grouping of dogs or cats.

§ 3.11 - Cleaning, sanitization, housekeeping, and pest control.

(a) *Cleaning of primary enclosures.* Excreta and food waste must be removed from primary enclosures daily, and from under primary enclosures as often as necessary to prevent an excessive accumulation of feces and food waste, to prevent soiling of the dogs or cats contained in the primary enclosures, and to reduce disease hazards, insects, pests and odors. When steam or water is used to clean the primary enclosure, whether by hosing, flushing, or other methods, dogs and cats must be removed, unless the enclosure is large enough to ensure the animals would not be harmed, wetted, or distressed in the process. Standing water must be removed from the primary enclosure and animals in other primary enclosures must be protected from being contaminated with water and other wastes during the cleaning. The pans under primary enclosures with grill-type floors and the ground areas under raised runs with mesh or slatted floors must be cleaned as often as necessary to prevent accumulation of feces and food waste and to reduce disease hazards pests, insects and odors.

(b) *Sanitization of primary enclosures and food and water receptacles.*

(1) Used primary enclosures and food and water receptacles must be cleaned and sanitized in accordance with this section before they can be used to house, feed, or water another dog or cat, or social grouping of dogs or cats.

(2) Used primary enclosures and food and water receptacles for dogs and cats must be sanitized at least once every 2 weeks using one of the methods prescribed in paragraph (b)(3) of this section, and more often if necessary to prevent an accumulation of dirt, debris, food waste, excreta, and other disease hazards.

(3) Hard surfaces of primary enclosures and food and water receptacles must be sanitized using one of the following methods:

(i) Live steam under pressure;

(ii) Washing with hot water (at least 180 °F (82.2 °C)) and soap or detergent, as with a mechanical cage washer; or

(iii) Washing all soiled surfaces with appropriate detergent solutions and disinfectants, or by using a combination detergent/disinfectant product that accomplishes the same purpose, with a thorough cleaning of the surfaces to remove organic material, so as to remove all organic material and mineral buildup, and to provide sanitization followed by a clean water rinse.

(4) Pens, runs, and outdoor housing areas using material that cannot be sanitized using the methods provided in paragraph (b)(3) of this section, such as gravel, sand, grass, earth, or absorbent bedding, must be sanitized by removing the contaminated material as necessary to prevent odors, diseases, pests, insects, and vermin infestation.

(c) *Housekeeping for premises.* Premises where housing facilities are located, including buildings and surrounding grounds, must be kept clean and in good repair to protect the animals from injury, to facilitate the husbandry practices required in this subpart, and to reduce or eliminate breeding and living areas for rodents and other pests and vermin. Premises must be kept free of accumulations of trash, junk, waste products, and discarded matter. Weeds, grasses, and bushes must be controlled so as to facilitate cleaning of the premises and pest control, and to protect the health and well-being of the animals.

(d) *Pest control.* An effective program for the control of insects, external parasites affecting dogs and cats, and birds and mammals that are pests, must be established and maintained so as to promote the health and well-being of the animals and reduce contamination by pests in animal areas.

[56 FR 6486, Feb. 15, 1991, as amended at 63 FR 3023, Jan. 21, 1998]

PART 3
Subpart A

124

§ 3.12 - Employees.

Each person subject to the Animal Welfare regulations (9 CFR parts 1, 2, and 3) maintaining dogs and cats must have enough employees to carry out the level of husbandry practices and care required in this subpart. The employees who provide for husbandry and care, or handle animals, must be supervised by an individual who has the knowledge, background, and experience in proper husbandry and care of dogs and cats to supervise others. The employer must be certain that the supervisor and other employees can perform to these standards.

TRANSPORTATION STANDARDS

§ 3.13 - Consignments to carriers and intermediate handlers.

(a) Carriers and intermediate handlers must not accept a dog or cat for transport in commerce more than 4 hours before the scheduled departure time of the primary conveyance on which the animal is to be transported. However, a carrier or intermediate handler may agree with anyone consigning a dog or cat to extend this time by up to 2 hours.

(b) Carriers and intermediate handlers must not accept a dog or cat for transport in commerce unless they are provided with the name, address, and telephone number of the consignee.

(c) Carriers and intermediate handlers must not accept a dog or cat for transport in commerce unless the consignor certifies in writing to the carrier or intermediate handler that the dog or cat was offered food and water during the 4 hours before delivery to the carrier or intermediate handler. The certification must be securely attached to the outside of the primary enclosure in a manner that makes it easily noticed and read. Instructions for no food or water are not acceptable unless directed by the attending veterinarian. Instructions must be in compliance with § 3.16 of this subpart. The certification must include the following information for each dog and cat:

(1) The consignor's name and address;

(2) The tag number or tattoo assigned to each dog or cat under §§ 2.38 and 2.50 of this chapter;

(3) The time and date the animal was last fed and watered and the specific instructions for the next feeding(s) and watering(s) for a 24-hour period; and

(4) The consignor's signature and the date and time the certification was signed.

(d) Carriers and intermediate handlers must not accept a dog or cat for transport in commerce in a primary enclosure unless the primary enclosure meets the requirements of § 3.14 of this subpart. A carrier or intermediate

PART 3
Subpart A

125

handler must not accept a dog or cat for transport if the primary enclosure is obviously defective or damaged and cannot reasonably be expected to safely and comfortably contain the dog or cat without causing suffering or injury.

(e) Carriers and intermediate handlers must not accept a dog or cat for transport in commerce unless their animal holding area meets the minimum temperature requirements provided in §§ 3.18 and 3.19 of this subpart, or unless the consignor provides them with a certificate signed by a veterinarian and dated no more than 10 days before delivery of the animal to the carrier or intermediate handler for transport in commerce, certifying that the animal is acclimated to temperatures lower than those required in §§ 3.18 and 3.19 of this subpart. Even if the carrier or intermediate handler receives this certification, the temperatures the dog or cat is exposed to while in a terminal facility must not be lower than 45 °F (2.2 °C) for more than 4 consecutive hours when dogs or cats are present, as set forth in § 3.18, nor lower than 45 °F (2.2 °C) for more than 45 minutes, as set forth in § 3.19, when moving dogs or cats to or from terminal facilities or primary conveyances. A copy of the certification must accompany the dog or cat to its destination and must include the following information:

PART 3

Subpart A

(1) The consignor's name and address;

(2) The tag number or tattoo assigned to each dog or cat under §§ 2.38 and 2.50 of this chapter;

(3) A statement by a veterinarian, dated no more than 10 days before delivery, that to the best of his or her knowledge, each of the dogs or cats contained in the primary enclosure is acclimated to air temperatures lower than 50 °F (10 °C); but not lower than a minimum temperature, specified on a certificate, that the attending veterinarian has determined is based on generally accepted temperature standards for the age, condition, and breed of the dog or cat; and

(4) The signature of the veterinarian and the date the certification was signed.

(f) When a primary enclosure containing a dog or cat has arrived at the animal holding area at a terminal facility after transport, the carrier or intermediate handler must attempt to notify the consignee upon arrival and at least once in every 6-hour period thereafter. The time, date, and method of all attempted notifications and the actual notification of the consignee, and the name of the person who notifies or attempts to notify the consignee must be written either on the carrier's or intermediate handler's copy of the shipping document or on the copy that accompanies the primary enclosure. If the consignee cannot be notified within 24 hours after the dog or cat has arrived at the terminal facility, the carrier or intermediate handler must return the animal to the consignor or to whomever the consignor designates. If the consignee is notified of the arrival and does not accept delivery of the dog or cat within 48 hours after arrival of the dog or cat, the carrier or intermediate handler must return the animal to the consignor or to whomever the consignor designates. The carrier or intermediate handler must continue to provide proper care, feeding, and housing to the dog or cat, and maintain the dog or cat in accordance with generally accepted professional and husbandry practices until the consignee accepts delivery of the dog or cat or until it is returned to the consignor or to whomever the consignor designates. The carrier or intermediate handler must obligate the consignor to reimburse the carrier or intermediate handler for the cost of return transportation and care.

(Approved by the Office of Management and Budget under control number 0579-0093)

§ 3.14 - Primary enclosures used to transport live dogs and cats.

Any person subject to the Animal Welfare regulations (9 CFR parts 1, 2, and 3) must not transport or deliver for transport in commerce a dog or cat unless the following requirements are met:

PART 3
Subpart A

(a) *Construction of primary enclosures.* The dog or cat must be contained in a primary enclosure such as a compartment, transport cage, carton, or crate. Primary enclosures used to transport dogs and cats must be constructed so that:

(1) The primary enclosure is strong enough to contain the dogs and cats securely and comfortably and to withstand the normal rigors of transportation;

(2) The interior of the primary enclosure has no sharp points or edges and no protrusions that could injure the animal contained in it;

(3) The dog or cat is at all times securely contained within the enclosure and cannot put any part of its body outside the enclosure in a way that could result in injury to itself, to handlers, or to persons or animals nearby;

(4) The dog or cat can be easily and quickly removed from the enclosure in an emergency;

(5) Unless the enclosure is permanently affixed to the conveyance, adequate devices such as handles or handholds are provided on its exterior, and enable the enclosure to be lifted without tilting it, and ensure that anyone handling the enclosure will not come into physical contact with the animal contained inside;

(6) Unless the enclosure is permanently affixed to the conveyance, it is clearly marked on top and on one or more sides with the words "Live Animals," in letters at least 1 inch (2.5 cm.) high, and with arrows or other markings to indicate the correct upright position of the primary enclosure;

(7) Any material, treatment, paint, preservative, or other chemical used in or on the enclosure is nontoxic to the animal and not harmful to the health or well-being of the animal;

(8) Proper ventilation is provided to the animal in accordance with paragraph (c) of this section; and

(9) The primary enclosure has a solid, leak-proof bottom or a removable, leak-proof collection tray under a slatted or mesh floor that prevents seepage of waste products, such as excreta and body fluids, outside of the enclosure. If a slatted or mesh floor is used in the enclosure, it must be designed and constructed so that the animal cannot put any part of its body between the slats or through the holes in the mesh. Unless the dogs and cats are on raised slatted floors or raised floors made of mesh, the primary enclosure must contain enough previously unused litter to absorb and cover excreta. The litter must be of a suitably absorbent material that is safe and nontoxic to the dogs and cats.

(b) *Cleaning of primary enclosures.* A primary enclosure used to hold or transport dogs or cats in commerce must be cleaned and sanitized before each use in accordance with the methods provided in § 3.11(b)(3) of this subpart. If the dogs or cats are in transit for more than 24 hours, the enclosures must be cleaned and any litter replaced, or other methods, such as moving the animals to another enclosure, must be utilized to prevent the soiling of the dogs or cats by body wastes. If it becomes necessary to remove the dog or cat from the enclosure in order to clean, or to move the dog or cat to another enclosure, this procedure must be completed in a way that safeguards the dog or cat from injury and prevents escape.

(c) *Ventilation.*

(1) Unless the primary enclosure is permanently affixed to the conveyance, there must be:

(i) Ventilation openings located on two opposing walls of the primary enclosure and the openings must be at least 16 percent of the surface area of each such wall, and the total combined surface area of the ventilation openings must be at least 14 percent of the total combined surface area of all the walls of the primary enclosure; or

(ii) Ventilation openings on three walls of the primary enclosure, and the openings on each of the two opposing walls must be at least 8 percent of the total surface area of the two walls, and the ventilation openings on the third wall of the primary enclosure must be at least 50 percent of the total surface area of that wall, and the total combined surface area of the ventilation openings must be at least 14 percent of the total combined surface area of all the walls of the primary enclosure; or

(iii) Ventilation openings located on all four walls of the primary enclosure and the ventilation openings on each of the four walls must be at least 8 percent of the total surface area of each such wall, and the total

128

combined surface area of the openings must be at least 14 percent of total combined surface area of all the walls of the primary enclosure; and

(iv) At least one-third of the ventilation area must be located on the upper half of the primary enclosure.

(2) Unless the primary enclosure is permanently affixed to the conveyance, projecting rims or similar devices must be located on the exterior of each enclosure wall having a ventilation opening, in order to prevent obstruction of the openings. The projecting rims or similar devices must be large enough to provide a minimum air circulation space of 0.75 in. (1.9 cm) between the primary enclosure and anything the enclosure is placed against.

(3) If a primary enclosure is permanently affixed to the primary conveyance so that there is only a front ventilation opening for the enclosure, the primary enclosure must be affixed to the primary conveyance in such a way that the front ventilation opening cannot be blocked, and the front ventilation opening must open directly to an unobstructed aisle or passageway inside the conveyance. The ventilation opening must be at least 90 percent of the total area of the front wall of the enclosure, and must be covered with bars, wire mesh, or smooth expanded metal having air spaces.

PART 3
Subpart A

(d) *Compatibility.*

(1) Live dogs or cats transported in the same primary enclosure must be of the same species and be maintained in compatible groups, except that dogs and cats that are private pets, are of comparable size, and are compatible, may be transported in the same primary enclosure.

(2) Puppies or kittens 4 months of age or less may not be transported in the same primary enclosure with adult dogs or cats other than their dams.

(3) Dogs or cats that are overly aggressive or exhibit a vicious disposition must be transported individually in a primary enclosure.

(4) Any female dog or cat in heat (estrus) may not be transported in the same primary enclosure with any male dog or cat.

(e) *Space and placement.*

(1) Primary enclosures used to transport live dogs and cats must be large enough to ensure that each animal contained in the primary enclosure has enough space to turn about normally while standing, to stand and sit erect, and to lie in a natural position.

(2) Primary enclosures used to transport dogs and cats must be positioned in the primary conveyance so as to provide protection from the elements.

(f) *Transportation by air.*

(1) No more than one live dog or cat, 6 months of age or older, may be transported in the same primary enclosure when shipped via air carrier.

(2) No more than one live puppy, 8 weeks to 6 months of age, and weighing over 20 lbs (9 kg), may be transported in a primary enclosure when shipped via air carrier.

(3) No more than two live puppies or kittens, 8 weeks to 6 months of age, that are of comparable size, and weighing 20 lbs (9 kg) or less each, may be transported in the same primary enclosure when shipped via air carrier.

(4) Weaned live puppies or kittens less than 8 weeks of age and of comparable size, or puppies or kittens that are less than 8 weeks of age that are littermates and are accompanied by their dam, may be transported in the same primary enclosure when shipped to research facilities, including Federal research facilities.

(g) *Transportation by surface vehicle or privately owned aircraft.*

(1) No more than four live dogs or cats, 8 weeks of age or older, that are of comparable size, may be transported in the same primary enclosure when shipped by surface vehicle (including ground and water transportation) or privately owned aircraft, and only if all other requirements of this section are met.

(2) Weaned live puppies or kittens less than 8 weeks of age and of comparable size, or puppies or kittens that are less than 8 weeks of age that are littermates and are accompanied by their dam, may be transported in the same primary enclosure when shipped to research facilities, including Federal research facilities, and only if all other requirements in this section are met.

(h) *Accompanying documents and records.* Shipping documents that must accompany shipments of dogs and cats may be held by the operator of the primary conveyance, for surface transportation only, or must be securely attached in a readily accessible manner to the outside of any primary enclosure that is part of the shipment, in a manner that allows them to be detached for examination and securely reattached, such as in a pocket or sleeve. Instructions for administration of drugs, medication, and other special care must be attached to each primary enclosure in a manner that makes them easy to notice, to detach for examination, and to reattach securely. Food and water instructions must be attached in accordance with § 3.13(c).

(Approved by the Office of Management and Budget under control number 0579-0093)

[56 FR 6486, Feb. 15, 1991, as amended at 63 FR 3023, Jan. 21, 1998]

§ 3.15 - Primary conveyances (motor vehicle, rail, air, and marine).

(a) The animal cargo space of primary conveyances used to transport dogs and cats must be designed, constructed, and maintained in a manner that at all times protects the health and well-being of the animals transported in them,

PART 3
Subpart A

ensures their safety and comfort, and prevents the entry of engine exhaust from the primary conveyance during transportation.

(b) The animal cargo space must have a supply of air that is sufficient for the normal breathing of all the animals being transported in it.

(c) Each primary enclosure containing dogs or cats must be positioned in the animal cargo space in a manner that provides protection from the elements and that allows each dog or cat enough air for normal breathing.

(d) During air transportation, dogs and cats must be held in cargo areas that are heated or cooled as necessary to maintain an ambient temperature and humidity that ensures the health and well-being of the dogs or cats. The cargo areas must be pressurized when the primary conveyance used for air transportation is not on the ground, unless flying under 8,000 ft. Dogs and cats must have adequate air for breathing at all times when being transported.

(e) During surface transportation, auxiliary ventilation, such as fans, blowers or air conditioning, must be used in any animal cargo space containing live dogs or cats when the ambient temperature within the animal cargo space reaches 85 °F (29.5 °C). Moreover, the ambient temperature may not exceed 85 °F (29.5 °C) for a period of more than 4 hours; nor fall below 45 °F (7.2 °C) for a period of more than 4 hours. The preceding requirements are in addition to, not in place of, all other requirements pertaining to climatic conditions in parts 2 and 3 of this chapter.

PART 3

Subpart A

(f) Primary enclosures must be positioned in the primary conveyance in a manner that allows the dogs and cats to be quickly and easily removed from the primary conveyance in an emergency.

(g) The interior of the animal cargo space must be kept clean.

(h) Live dogs and cats may not be transported with any material, substance (e.g., dry ice) or device in a manner that may reasonably be expected to harm the dogs and cats or cause inhumane conditions.

[56 FR 6486, Feb. 15, 1991, as amended at 63 FR 10498, 10499, Mar. 4, 1998]

§ 3.16 - Food and water requirements.

(a) Each dog and cat that is 16 weeks of age or more must be offered food at least once every 24 hours. Puppies and kittens less than 16 weeks of age must be offered food at least once every 12 hours. Each dog and cat must be offered potable water at least once every 12 hours. These time periods apply to dealers, exhibitors, research facilities. including Federal research facilities, who transport dogs and cats in their own primary conveyance, starting from the time the dog or cat was last offered food and potable water before transportation was begun. These time periods apply to carriers and intermediate handlers starting from the date and time stated on the certificate provided under § 3.13(c) of this subpart. Each dog and cat must be offered food and potable water within 4 hours before being transported in commerce. Consignors who are subject to the Animal Welfare regulations (9 CFR parts 1, 2, and 3) must certify that each dog and cat was offered food and potable water within the 4 hours preceding delivery of the dog or cat to a carrier or intermediate handler for transportation in commerce, and must certify the date and time the food and potable water was offered, in accordance with § 3.13(c) of this subpart.

(b) Any dealer, research facility, including a Federal research facility, or exhibitor offering any dog or cat to a carrier or intermediate handler for transportation in commerce must securely attach to the outside of the primary enclosure used for transporting the dog or cat, written instructions for the in-transit food and water requirements for a 24-hour period for the dogs and cats contained in the enclosure. The instructions must be attached in a manner that makes them easily noticed and read.

(c) Food and water receptacles must be securely attached inside the primary enclosure and placed so that the receptacles can be filled from outside the enclosure without opening the door. Food and water containers must be designed, constructed, and installed so that a dog or cat cannot leave the primary enclosure through the food or water opening.

(Approved by the Office of Management and Budget under control number 0579-0093)

§ 3.17 - Care in transit.

(a) *Surface transportation (ground and water).* Any person subject to the Animal Welfare regulations transporting dogs or cats in commerce must ensure that the operator of the conveyance, or a person accompanying the operator, observes the dogs or cats as often as circumstances allow, but not less than once every 4 hours, to make sure they have sufficient air for normal breathing, that the ambient temperature is within the limits provided in § 3.15(e), and that all applicable standards of this subpart are being complied with. The regulated person must ensure that the operator or person accompanying the operator determines whether any of the dogs or cats are in obvious physical distress and obtains any veterinary care needed for the dogs or cats at the closest available veterinary facility.

(b) *Air transportation.* During air transportation of dogs or cats, it is the responsibility of the carrier to observe the dogs or cats as frequently as circumstances allow, but not less than once every 4 hours if the animal cargo area is accessible during flight. If the animal cargo area is not accessible during flight, the carrier must observe the dogs or cats whenever they are loaded and unloaded and whenever the animal cargo space is otherwise accessible to make sure they have sufficient air for normal breathing, that the animal cargo area meets the heating and cooling requirements of § 3.15(d), and that all other applicable standards of this subpart are being complied with. The carrier must determine whether any of the dogs or cats are in obvious physical distress, and arrange for any needed veterinary care as soon as possible.

(c) If a dog or cat is obviously ill, injured, or in physical distress, it must not be transported in commerce, except to receive veterinary care for the condition.

(d) Except during the cleaning of primary enclosures, as required in § 3.14(b) of this subpart, during transportation in commerce a dog or cat must not be removed from its primary enclosure, unless it is placed in another primary enclosure or facility that meets the requirements of § 3.6 or § 3.14 of this subpart.

(e) The transportation regulations contained in this subpart must be complied with until a consignee takes physical delivery of the dog or cat if the animal is consigned for transportation, or until the animal is returned to the consignor.

§ 3.18 - Terminal facilities.

(a) *Placement.* Any person subject to the Animal Welfare regulations (9 CFR parts 1, 2, and 3) must not commingle shipments of dogs or cats with inanimate cargo in animal holding areas of terminal facilities.

(b) *Cleaning, sanitization, and pest control.* All animal holding areas of terminal facilities must be cleaned and sanitized in a manner prescribed in § 3.11(b)(3) of this subpart, as often as necessary to prevent an accumulation of debris or excreta and to minimize vermin infestation and disease hazards. Terminal facilities must follow an effective program in all animal holding areas for the control of insects, ectoparasites, and birds and mammals that are pests to dogs and cats.

(c) *Ventilation.* Ventilation must be provided in any animal holding area in a terminal facility containing dogs or cats, by means of windows, doors, vents, or air conditioning. The air must be circulated by fans, blowers, or air conditioning so as to minimize drafts, odors, and moisture condensation. Auxiliary ventilation, such as exhaust fans, vents, fans, blowers, or air conditioning must be used in any animal holding area containing dogs and cats, when the ambient temperature is 85 °F (29.5 °C) or higher

(d) *Temperature.* The ambient temperature in an animal holding area containing dogs or cats must not fall below 45 °F (7.2 °C) or rise above 85 °F (29.5 °C) for more than four consecutive hours at any time dogs or cats are present. The ambient temperature must be measured in the animal holding area by the carrier, intermediate handler, or a person transporting dogs or cats who is subject to the Animal Welfare regulations (9 CFR parts 1, 2, and 3), outside any primary enclosure containing a dog or cat at a point not more than 3 feet (0.91 m) away from an outside wall of the primary enclosure, and approximately midway up the side of the enclosure. The preceding requirements are in addition to, not in place of, all other requirements pertaining to climatic conditions in parts 2 and 3 of this chapter.

(e) *Shelter.* Any person subject to the Animal Welfare regulations (9 CFR parts 1, 2, and 3) holding a live dog or cat in an animal holding area of a terminal facility must provide the following:

(1) *Shelter from sunlight and extreme heat.* Shade must be provided that is sufficient to protect the dog or cat from the direct rays of the sun.

(2) *Shelter from rain or snow.* Sufficient protection must be provided to allow the dogs and cats to remain dry during rain, snow, and other precipitation.

(f) *Duration.* The length of time any person subject to the Animal Welfare regulations (9 CFR parts 1, 2, and 3) can hold dogs and cats in animal holding areas of terminal facilities upon arrival is the same as that provided in § 3.13(f) of this subpart.

[56 FR 6486, Feb. 15, 1991, as amended at 63 FR 10499, Mar. 4, 1998]

PART 3
Subpart A

§ 3.19 - Handling.

(a) Any person subject to the Animal Welfare regulations (9 CFR parts 1, 2, and 3) who moves (including loading and unloading) dogs or cats within, to, or from the animal holding area of a terminal facility or a primary conveyance must do so as quickly and efficiently as possible and must provide the following during movement of the dog or cat:

(1) *Shelter from sunlight and extreme heat.* Sufficient shade must be provided to protect the dog or cat from the direct rays of the sun. The dog or cat must not be exposed to an ambient air temperature above 85 °F (29.5 °C) for a period of more than 45 minutes while being moved to or from a primary conveyance or a terminal facility. The temperature must be measured in the manner provided in § 3.18(d) of this subpart. The preceding requirements are in addition to, not in place of, all other requirements pertaining to climatic conditions in parts 2 and 3 of this chapter.

(2) *Shelter from rain and snow.* Sufficient protection must be provided to allow the dogs and cats to remain dry during rain, snow, and other precipitation.

(3) *Shelter from cold temperatures.* Transporting devices on which live dogs or cats are placed to move them must be covered to protect the animals when the outdoor temperature falls below 50 °F (10 °C). The dogs or cats must not be exposed to an ambient temperature below 45 °F (7.2

°C) for a period of more than 45 minutes, unless they are accompanied by a certificate of acclimation to lower temperatures as provided in § 3.13(e). The temperature must be measured in the manner provided in § 3.18(d) of this subpart. The preceding requirements are in addition to, not in place of, all other requirements pertaining to climatic conditions in parts 2 and 3 of this chapter.

(b) Any person handling a primary enclosure containing a dog or cat must use care and must avoid causing physical harm or distress to the dog or cat.

(1) A primary enclosure containing a live dog or cat must not be placed on unattended conveyor belts, or on elevated conveyor belts, such as baggage claim conveyor belts and inclined conveyor ramps that lead to baggage claim areas, at any time; except that a primary enclosure may be placed on inclined conveyor ramps used to load and unload aircraft if an attendant is present at each end of the conveyor belt.

(2) A primary enclosure containing a dog or cat must not be tossed, dropped, or needlessly tilted, and must not be stacked in a manner that may reasonably be expected to result in its falling. It must be handled and positioned in the manner that written instructions and arrows on the outside of the primary enclosure indicate.

PART 3
Subpart A

(c) This section applies to movement of a dog or cat from primary conveyance to primary conveyance, within a primary conveyance or terminal facility, and to or from a terminal facility or a primary conveyance.

(Approved by the Office of Management and Budget under control number 0579-0093)

[56 FR 6486, Feb. 15, 1991, as amended at 63 FR 10499, Mar. 4, 1998]

Subpart B – Specifications for the Humane Handling, Care, Treatment, and Transportation of Guinea Pigs and Hamsters

FACILITIES AND OPERATING STANDARDS

§ 3.25 - Facilities, general.

(a) *Structural strength.* Indoor and outdoor housing facilities for guinea pigs or hamsters shall be structurally sound and shall be maintained in good repair, to protect the animals from injury, to contain the animals, and to restrict the entrance of other animals.

(b) *Water and electric power.* Reliable and adequate electric power, if required to comply with other provisions of this subpart, and adequate potable water shall be available.

(c) *Storage.* Supplies of food and bedding shall be stored in facilities which adequately protect such supplies against spoilage or deterioration and infestation or contamination by vermin. Food supplies shall be stored in containers with tightly fitting lids or covers or in the original containers as received from the commercial sources of supply. Refrigeration shall be provided for supplies of perishable food.

(d) *Waste disposal.* Provisions shall be made for the removal and disposal of animal and food wastes, bedding, dead animals, and debris. Disposal facilities shall be so provided and operated as to minimize vermin infestation, odors, and disease hazards.

(e) *Washroom and sinks.* Facilities, such as washrooms, basins, or sinks, shall be provided to maintain cleanliness among animal caretakers.

PART 3
Subpart B

[32 FR 3273, Feb. 24, 1967, as amended at 44 FR 63492, Nov. 2, 1979]

§ 3.26 - Facilities, indoor.

(a) *Heating.* Indoor housing facilities for guinea pigs or hamsters shall be sufficiently heated when necessary to protect the animals from the cold, and to provide for their health and comfort. The ambient temperature shall not be allowed to fall below 60 °F. nor to exceed 85 °F.

(b) *Ventilation.* Indoor housing facilities for guinea pigs or hamsters shall be adequately ventilated to provide for the health and comfort of the animals at all times. Such facilities shall be provided with fresh air either by means of windows, doors, vents, or air conditioning, and shall be ventilated so as to minimize drafts, odors, and moisture condensation. The ambient temperature shall not be allowed to rise above 85 °F.

(c) *Lighting.* Indoor housing facilities for guinea pigs or hamsters shall have ample light, by natural or artificial means, or both, of good quality and well distributed. Such lighting shall provide uniformly distributed illumination of sufficient light intensity to permit routine inspection and cleaning during the entire working period. Primary enclosures shall be so placed as to protect the guinea pigs or hamsters from excessive illumination.

(d) *Interior surfaces.* The interior building surfaces of indoor housing facilities shall be constructed and maintained so that they are substantially impervious to moisture and may be readily sanitized.

§ 3.27 - Facilities, outdoor.

(a) Hamsters shall not be housed in outdoor facilities.

(b) Guinea pigs shall not be housed in outdoor facilities unless such facilities are located in an appropriate climate and prior approval for such outdoor housing is obtained from the Deputy Administrator.

§ 3.28 - Primary enclosures.

All primary enclosures for guinea pigs and hamsters shall conform to the following requirements:

(a) *General.*

(1) Primary enclosures shall be structurally sound and maintained in good repair to protect the guinea pigs and hamsters from injury. Such enclosures, including their racks, shelving and other accessories, shall be constructed of smooth material substantially impervious to liquids and moisture.

(2) Primary enclosures shall be constructed and maintained so that the guinea pigs or hamsters contained therein have convenient access to clean food and water as required in this subpart.

(3) Primary enclosures having a solid floor shall be provided with clean bedding material.

(4) Primary enclosures equipped with mesh or wire floors shall be so constructed as to allow feces to pass through the spaces of the mesh or wire: *Provided, however,* That such floors shall be constructed so as to protect the animals' feet and legs from injury.

(b) *Space requirements for primary enclosures acquired before August 15, 1990*

(1) *Guinea pigs and hamsters.* Primary enclosures shall be constructed and maintained so as to provide sufficient space for each animal contained therein to make normal postural adjustments with adequate freedom of movement.

(2) *Guinea pigs.* In addition to the provisions of paragraph (b)(1) of this section, the following space requirements are applicable to primary enclosures for guinea pigs:

(i) The interior height of any primary enclosure used to confine guinea pigs shall be at least 6½ inches.

138

(ii) Each guinea pig housed in a primary enclosure shall be provided a minimum amount of floor space in accordance with the following table:

Weight or stage of maturity	Minimum space per guinea pig (square inches)
Weaning to 350 grams	60
350 grams or more	90
Breeders	180

(3) *Hamsters.* In addition to the provisions of paragraph (b)(1) of this section, the following space requirements are applicable to primary enclosures for hamsters:

(i) The interior height of any primary enclosure used to confine hamsters shall be at least 5½ inches, except that in the case of dwarf hamsters, such interior height shall be at least 5 inches.

(ii) A nursing female hamster, together with her litter, shall be housed in a primary enclosure which contains no other hamsters and which provides at least 121 square inches of floor space: *Provided, however,* That in the case of dwarf hamsters such floor space shall be at least 25 square inches.

(iii) The minimum amount of floor space per individual hamster and the maximum number of hamsters allowed in a single primary enclosure, except as provided for nursing females in paragraph (b)(3)(ii) of this section, shall be in accordance with the following table:

PART 3
Subpart B

Age	Minimum space per hamster (square inches)		Maximum population per enclosure
	Dwarf	Other	
Weaning to 5 wks	5.0	10.0	20
5 to 10 wks	7.5	12.5	16
10 wks or more	9	15.0	13

(c) *Space requirements for primary enclosures acquired on or after August 15, 1990*

(1) *Guinea pigs.*

(i) Primary enclosures shall be constructed and maintained so as to provide sufficient space for each guinea pig contained therein to make normal postural adjustments with adequate freedom of movement.

(ii) The interior height of any primary enclosure used to confine guinea pigs shall be at least 7 inches (17.78 cm).

(iii) Each guinea pig shall be provided a minimum amount of floor space in any primary enclosure as follows:

Weight or stage of maturity	Minimum floor space	
	in²	cm²
Weaning to 350 grams	60	387.12
> 350 grams	101	651.65
Nursing females with their litters	101	651.65

(2) Hamsters.

(i) Primary enclosures shall be constructed and maintained so as to provide sufficient space for each hamster contained therein to make normal postural adjustments with adequate freedom of movement.

(ii) The interior height of any primary enclosure used to confine hamsters shall be at least 6 inches (15.24 cm).

(iii) Except as provided in paragraph (c)(2)(iv) of this section, each hamster shall be provided a minimum amount of floor space in any primary enclosure as follows:

PART 3

Subpart B

Weight		Minimum floor space per hamster	
G	ozs	in²	cm²
< 60	< 2.1	10	64.52
60 to 80	2.1-2.8	13	83.88
80 to 100	2.8-3.5	16	103.23
> 100	> 3.5	19	122.59

(iv) A nursing female hamster, together with her litter, shall be housed in a primary enclosure that contains no other hamsters and that provides at least 121 square inches of floor space: *Provided, however,* That in the case of nursing female dwarf hamsters such floor space shall be at least 25 square inches.

(3) Innovative primary enclosures that do not precisely meet the space requirements of paragraph (c)(1) or (c)(2) of this section, but that do provide guinea pigs or hamsters with a sufficient volume of space and the opportunity to express species-typical behavior, may be used at research facilities when

approved by the Institutional Animal Care and Use Committee, and by dealers and exhibitors when approved by the Administrator.

[32 FR 3273, Feb. 24, 1967, as amended at 55 FR 28882, July 16, 1990]

ANIMAL HEALTH AND HUSBANDRY STANDARDS

§ 3.29 - Feeding.

(a) Guinea pigs and hamsters shall be fed each day except as otherwise might be required to provide adequate veterinary care. The food shall be free from contamination, wholesome, palatable and of sufficient quantity and nutritive value to meet the normal daily requirements for the condition and size of the guinea pig or hamster.

(b) Food comprising the basic diet shall be at least equivalent in quality and content to pelleted rations produced commercially and commonly available from feed suppliers.

(c) The basic diet of guinea pigs and hamsters may be supplemented with good quality fruits or vegetables consistent with their individual dietary requirements.

(d) Food receptacles, if used, shall be accessible to all guinea pigs or hamsters in a primary enclosure and shall be located so as to minimize contamination by excreta. All food receptacles shall be kept clean and shall be sanitized at least once every 2 weeks. If self-feeders are used for the feeding of pelleted feed, measures must be taken to prevent molding, deterioration or caking of the feed. Hamsters may be fed pelleted feed on the floor of a primary enclosure.

(e) Fruit or vegetable food supplements may be placed upon the bedding within the primary enclosure: *Provided, however,* That the uneaten portion of such supplements and any bedding soiled as a result of such feeding practices shall be removed from the primary enclosure when such uneaten supplements accumulate or such bedding becomes soiled to a degree that might be harmful or uncomfortable to animals therein.

PART 3
Subpart B

§ 3.30 - Watering.

Unless food supplements consumed by guinea pigs or hamsters supply them with their normal water requirements, potable water shall be provided daily except as might otherwise be required to provide adequate veterinary care. Open containers used for dispensing water to guinea pigs or hamsters shall be so placed in or attached to the primary enclosure as to minimize contamination from excreta. All watering receptacles shall be sanitized when dirty: *Provided, however,* That such receptacles shall be sanitized at least once every 2 weeks.

§ 3.31 - Sanitation.

(a) *Cleaning and sanitation of primary enclosures.*

(1) Primary enclosures shall be cleaned and sanitized often enough to prevent an accumulation of excreta or debris: *Provided, however,* That such enclosures shall be sanitized at least once every 2 weeks in the manner provided in paragraph (a)(4) of this section.

(2) In the event a primary enclosure becomes soiled or wet to a degree that might be harmful or uncomfortable to the animals therein due to leakage of the watering system, discharges from dead or dying animals, spoiled perishable foods, or moisture condensation, the guinea pigs or hamsters shall be transferred to clean primary enclosures.

(3) Prior to the introduction of guinea pigs or hamsters into empty primary enclosures previously occupied, such enclosures shall be sanitized in the manner provided in paragraph (a)(4) of this section.

(4) Primary enclosures for guinea pigs or hamsters shall be sanitized by washing them with hot water (180 °F.) and soap or detergent as in a mechanical cage washer, or by washing all soiled surfaces with a detergent solution followed by a safe and effective disinfectant, or by cleaning all soiled surfaces with live steam.

(b) *Housekeeping.* Premises (buildings and grounds) shall be kept clean and in good repair in order to protect the animals from injury and to facilitate the prescribed husbandry practices set forth in this subpart. Premises shall remain free of accumulations of trash.

(c) *Pest control.* An effective program for the control of insects, ectoparasites, and avian and mammalian pests shall be established and maintained.

§ 3.32 - Employees.

A sufficient number of employees shall be utilized to maintain the prescribed level of husbandry practices set forth in this subpart. Such practices shall be under the supervision of an animal caretaker who has a background in animal husbandry or care.

§ 3.33 - Classification and separation.

Animals housed in the same primary enclosure shall be maintained in compatible groups, with the following additional restrictions:

(a) Except where harem breeding is practiced, preweanling guinea pigs shall not be housed in the same primary enclosure with adults other than their parents.

(b) Guinea pigs shall not be housed in the same primary enclosure with hamsters, nor shall guinea pigs or hamsters be housed in the same primary enclosure with any other species of animals.

(c) Guinea pigs or hamsters under quarantine or treatment for a communicable disease shall be separated from other guinea pigs or hamsters and other susceptible species of animals in such a manner as to minimize dissemination of such disease.

§ 3.34 - [Reserved]

TRANSPORTATION STANDARDS

Authority: Sections 3.35 through 3.41 issued under secs. 3, 5, 6, 10, 11, 14, 16, 17, 21; 80 Stat. 353; 84 Stat. 1561, 1562, 1563, 1564; 90 Stat. 418, 419, 420, 423; (7 U.S.C. 2133, 2135, 2136, 2140, 2141, 2144, 2146, 2147, 2151); 37 FR 28464, 28477, 38 FR 19141.

§ 3.35 - Consignments to carriers and intermediate handlers.

(a) Carriers and intermediate handlers shall not accept any live guinea pig or hamster presented by any dealer, research facility, exhibitor, operator of an auction sale, or other person, or any department, agency, or instrumentality of the United States or any State or local government for shipment, in commerce, more than 4 hours prior to the scheduled departure of the primary conveyance on which it is to be transported: *Provided, however,* That the carrier or intermediate handler and any dealer, research facility, exhibitor, operator of an auction sale, or other person, or any department, agency, or instrumentality of the United States or any State or local government may mutually agree to extend the time of acceptance to not more than 6 hours if specific prior scheduling of the animal shipment to destination has been made.

(b) Any carrier or intermediate handler shall only accept for transportation or transport, in commerce any live guinea pig or hamster in a primary enclosure which conforms to the requirements set forth in § 3.36 of the standards: *Provided, however,* That any carrier or intermediate handler may accept for transportation or transport, in commerce, any live guinea pig or hamster consigned by any department, agency, or instrumentality of the United States having laboratory animal facilities or exhibiting animals, or any licensed or registered dealer, research facility, exhibitor, or operator of an auction sale, if such consignor furnishes to the carrier or intermediate handler a certificate, signed by the consignor, stating that the primary enclosure complies with § 3.36 of the standards, unless such primary enclosure is obviously defective or damaged and it is apparent that it cannot reasonably be

PART 3
Subpart B

expected to contain the live guinea pig or hamster without causing suffering or injury to such live guinea pig or hamster. A copy of such certificate shall accompany the shipment to destination. The certificate of compliance shall include at least the following information:

(1) Name and address of the consignor;

(2) The number of guinea pigs or hamsters in the primary enclosure(s);

(3) A certifying statement (e.g., "I hereby certify that the __ (number) primary enclosure(s) which are used to transport the animal(s) in this shipment complies (comply) with USDA standards for primary enclosures (9 CFR part 3)."); and

(4) The signature of the consignor, and date.

(c) Carriers or intermediate handlers whose facilities fail to meet the minimum temperature allowed by the standards may accept for transportation or transport, in commerce, any live hamster consigned by any department, agency, or instrumentality of the United States or of any State or local government, or by any person (including any licensee or registrant under the Act, as well as any private individual) if the consignor furnishes to the carrier or intermediate handler a certificate executed by a veterinarian accredited by this Department pursuant to part 160 of this title on a specified date which shall not be more than 10 days prior to delivery of such hamster for transportation in commerce, stating that such live hamster is acclimated to air temperatures lower than those prescribed in §§ 3.40 and 3.41. A copy of such certificate shall accompany the shipment to destination. The certificate shall include the following information:

(1) Name and address of the consignor;

(2) The number of hamsters in the shipment;

(3) A certifying statement (e.g., "I hereby certify that the animal(s) in this shipment is (are), to the best of my knowledge, acclimated to air temperatures lower than 7.2 °C. (45 °F.)."); and

(4) The signature of the USDA accredited veterinarian, assigned accreditation number, and date.

(d) Carriers and intermediate handlers shall attempt to notify the consignee at least once in every 6 hour period following the arrival of any live guinea pig or hamster at the animal holding area of the terminal cargo facility. The time, date, and method of each attempted notification and the final notification to the consignee and the name of the person notifying the consignee shall be recorded on the copy of the shipping document retained by the carrier or intermediate handler and on a copy of the shipping document accompanying the animal shipment.

[42 FR 31563, June 21, 1977, as amended at 43 FR 22163, May 16, 1978; 44 FR 63492, Nov. 2, 1979]

§3.36 - Primary enclosures used to transport live guinea pigs and hamsters.

No person subject to the Animal Welfare regulations shall offer for transportation, or transport, in commerce any live guinea pig or hamster in a primary enclosure that does not conform to the following requirements:

(a) Primary enclosures, such as compartments, transport cages, cartons, or crates, used to transport live guinea pigs or hamsters shall be constructed in such a manner that **(1)** the structural strength of the enclosure shall be sufficient to contain the live guinea pigs or hamsters and to withstand the normal rigors of transportation; **(2)** the interior of the enclosure shall be free from any protrusions that could be injurious to the live guinea pigs or hamsters contained therein; **(3)** the inner surfaces of corrugated fiberboard, cardboard, or plastic containers shall be covered or laminated with wire mesh or screen where necessary to prevent escape of the animals; **(4)** the openings of such enclosures are easily accessible at all times for emergency removal of the live guinea pigs or hamsters; **(5)** except as provided in paragraph (i) of this section, there are ventilation openings located on two opposite walls of the primary enclosure and the ventilation openings on each such wall shall be at least 16 percent of the total surface area of each such wall, or there are ventilation openings located on all four walls of the primary enclosure and the ventilation openings on each such wall shall be at least 8 percent of the total surface area of each such wall: *Provided, however,* That at least one-third of the total minimum area required for ventilation of the primary enclosure shall be located on the lower one-half of the primary enclosure and at least one-third of the total minimum area required for ventilation of the primary enclosure shall be located on the upper one-half of the primary enclosure; **(6)** except as provided in paragraph (i) of this section, projecting rims or other devices shall be on the exterior of the outside walls with any ventilation openings to prevent obstruction of the ventilation openings and to provide a minimum air circulation space of 1.9 centimeters (0.75 inches) between the primary enclosure and any adjacent cargo or conveyance wall; and **(7)** except as provided in paragraph (i) of this section, adequate handholds or other devices for lifting shall be provided on the exterior of the primary enclosure to enable the primary enclosure to be lifted without tilting and to ensure that the person handling the primary enclosure will not be in contact with the guinea pigs or hamsters.

(b) Live guinea pigs or hamsters transported in the same primary enclosure shall be of the same species and maintained in compatible groups.

(c) Primary enclosures used to transport live guinea pigs or hamsters shall be large enough to ensure that each animal contained therein has sufficient space to turn about freely and to make normal postural adjustments.

PART 3
Subpart B

(d) Not more than 15 live guinea pigs shall be transported in the same primary enclosure. No more than 50 live hamsters shall be transported in the same primary enclosure.

(e) In addition to the other provisions of this section, the following requirements shall also apply to primary enclosures used to transport live guinea pigs or hamsters:

(1) *Guinea pigs.*

(i) The interior height of primary enclosures used to transport live guinea pigs weighing up to 500 grams shall be at least 15.2 centimeters (6 inches) and the interior height of primary enclosures used to transport live guinea pigs weighing over 500 grams shall be at least 17.8 centimeters (7 inches).

(ii) Each live guinea pig transported in a primary enclosure shall be provided a minimum amount of floor space in accordance with the following table:

MINIMUM SPACE PER LIVE GUINEA PIG

Weight (grams)	Square centimeters	Square inches
Up to 350	193.6	30
350 to 600	290.3	45
Over 600	354.8	55

PART 3
Subpart B

(2) *Hamsters.*

(i) The interior height of primary enclosures used to transport live hamsters shall be at least 15.2 centimeters (6 inches) except that in the case of dwarf hamsters such interior height shall be at least 12.7 centimeters (5 inches).

(ii) Each live hamster transported in a primary enclosure shall be provided a minimum amount of floor space in accordance with the following table:

MINIMUM SPACE PER LIVE HAMSTER

Age	Dwarf		Other	
	Square centimeters	Square inches	Square centimeters	Square inches
Weaning to 5 wks	32.2	5.0	45.2	7
5 to 10 wks	48.3	7.5	71.0	11
Over 10 wks	58.1	9.0	96.8	15

(f) Primary enclosures used to transport live guinea pigs or hamsters as provided in this section shall have solid bottoms to prevent leakage in shipment and shall be cleaned and sanitized in a manner prescribed in § 3.31 of the standards, if previously used. Such primary enclosures shall contain clean litter of a suitable absorbent material, which is safe and nontoxic to the guinea pigs or hamsters, in sufficient quantity to absorb and cover excreta, unless the guinea pigs or hamsters are on wire or other nonsolid floors.

(g) Primary enclosures used to transport live guinea pigs or hamsters, except where such primary enclosures are permanently affixed in the animal cargo space of the primary conveyance, shall be clearly marked on top and on one or more sides with the words "Live Animals" in letters not less than 2.5 centimeters (1 inch) in height, and with arrows or other markings, to indicate the correct upright position of the container.

PART 3
Subpart B

(h) Documents accompanying the shipment shall be attached in an easily accessible manner to the outside of a primary enclosure which is part of such shipment.

(i) When a primary enclosure is permanently affixed within the animal cargo space of the primary conveyance so that the front opening is the only source of ventilation for such primary enclosure, the front opening shall open directly to the outside or to an unobstructed aisle or passageway within the primary conveyance. Such front ventilation opening shall be at least 90 percent of the total surface area of the front wall of the primary enclosure and covered with bars, wire mesh or smooth expanded metal.

[42 FR 31563, June 21, 1977, as amended at 43 FR 21163, May 16, 1978; 55 FR 28882, July 16, 1990]

§ 3.37 - Primary conveyances (motor vehicle, rail, air, and marine).

(a) The animal cargo space of primary conveyances used in transporting live guinea pigs and hamsters shall be designed and constructed to protect the health, and ensure the safety and comfort of the live guinea pigs and hamsters at all times.

(b) The animal cargo space shall be constructed and maintained in a manner to prevent the ingress of engine exhaust fumes and gases from the primary conveyance during transportation in commerce.

(c) No live guinea pig or hamster shall be placed in an animal cargo space that does not have a supply of air sufficient for normal breathing for each live animal contained therein, and the primary enclosures shall be positioned in the animal cargo space in such a manner that each live guinea pig or hamster has access to sufficient air for normal breathing.

(d) Primary enclosures shall be positioned in the primary conveyance in such a manner that in an emergency the live guinea pigs or hamsters can be removed from the primary conveyance as soon as possible.

(e) The interior of the animal cargo space shall be kept clean.

(f) Live guinea pigs and hamsters shall not be transported with any material, substance (e.g., dry ice) or device which may reasonably be expected to be injurious to the health and well-being of the guinea pigs and hamsters unless proper precaution is taken to prevent such injury.

(g) The animal cargo space of primary conveyances used to transport guinea pigs or hamsters shall be mechanically sound and provide fresh air by means of windows, doors, vents, or air conditioning so as to minimize drafts, odors, and moisture condensation. Auxiliary ventilation, such as fans, blowers, or air conditioners, shall be used in any cargo space containing live guinea pigs or hamsters when the ambient temperature in the animal cargo space is 75 °F (23.9 °C) or higher. The ambient temperature within the animal cargo space shall not exceed 85 °F (29.5 °C) or fall below 45 °F (7.2 °C), except that the ambient temperature in the cargo space may be below 45 °F (7.2 °C) for hamsters if the hamsters are accompanied by a certificate of acclimation to lower temperatures, as provided in § 3.35(c) of this part.

[42 FR 31563, June 21, 1977, as amended at 55 FR 28882, July 16, 1990]

§ 3.38 - Food and water requirements.

(a) If live guinea pigs or hamsters are to be transported for a period of more than 6 hours, the animals shall have access to food and water or a type of food, which provides the requirements for food and water in quantity and quality sufficient to satisfy their food and water needs, during transit.

(b) Any dealer, research facility, exhibitor or operator of an auction sale offering any live guinea pig or hamster to any carrier or intermediate handler for transportation, in commerce, shall provide an adequate supply of food or type of food, which provides the requirements for food and water, within the primary enclosure to meet the requirements of this section.

(c) No carrier or intermediate handler shall accept for transportation, in commerce, any live guinea pig or hamster without an adequate supply of food or type of food, which provides the requirements for food and water, within the primary enclosure to meet the requirements of this section.

[42 FR 31563, June 21, 1977]

§ 3.39 - Care in transit.

(a) During surface transportation, it shall be the responsibility of the driver or other employee to visually observe the live guinea pigs or hamsters as frequently as circumstances may dictate, but not less than once every 4 hours, to assure that they are receiving sufficient air for normal breathing, their ambient temperatures are within the prescribed limits, all other applicable standards are being complied with and to determine whether any of the live guinea pigs or hamsters are in obvious physical distress and to provide any needed veterinary care as soon as possible. When transported by air, live guinea pigs and hamsters shall be visually observed by the carrier as frequently as circumstances may dictate, but not less than once every 4 hours, if the animal cargo space is accessible during flight. If the animal cargo space is not accessible during flight, the carrier shall visually observe the live guinea pigs or hamsters whenever loaded and unloaded and whenever the animal cargo space is otherwise accessible to assure that they are receiving sufficient air for normal breathing, their ambient temperatures are within the prescribed limits, all other applicable standards are being complied with and to determine whether any such live guinea pigs or hamsters are in obvious physical distress. The carrier shall provide any needed veterinary care as soon as possible. No guinea pig or hamster in obvious physical distress shall be transported in commerce.

(b) During the course of transportation, in commerce, live guinea pigs or hamsters shall not be removed from their primary enclosures unless placed in other primary enclosures or facilities conforming to the requirements provided in this subpart.

PART 3
Subpart B

[42 FR 31563, June 21, 1977]

§ 3.40 - Terminal facilities.

No person subject to the Animal Welfare regulations shall commingle shipments of live guinea pigs or hamsters with inanimate cargo. All animal holding areas of a terminal facility where shipments of live guinea pigs or hamsters are maintained shall be cleaned and sanitized as prescribed in § 3.31 of the standards often enough to prevent an accumulation of debris or excreta, to minimize vermin infestation, and to prevent a disease hazard. An effective program for the control of insects, ectoparasites, and avian and mammalian pests shall be established and maintained for all animal holding areas. Any animal holding area containing live guinea pigs or hamsters shall be provided with fresh air by means of windows, doors, vents, or air conditioning and may be ventilated or air circulated by means of fans, blowers, or an air conditioning system so as to minimize drafts, odors, and moisture condensation. Auxiliary ventilation, such as exhaust fans and vents or fans or blowers or air conditioning shall be used for any animal holding area containing live guinea pigs and hamsters when the air temperature within such animal holding area is 23.9 °C. (75. °F.) or higher. The air temperature around any live guinea pig or hamster in any animal holding area shall not be allowed to fall below 7.2 °C. (45 °F.) nor be allowed to exceed 29.5 °C. (85 °F.) at any time. To ascertain compliance with the provisions of this paragraph, the air temperature around any live guinea pig or hamster shall be measured and read outside the primary enclosure which contains such guinea pig or hamster at a distance not to exceed .91 meters (3 feet) from any one of the external walls of the primary enclosure and measured on a level parallel to the bottom of such primary enclosure at a point which approximates half the distance between the top and bottom of such primary enclosure.

PART 3
Subpart B

[43 FR 56215, Dec. 1, 1978, as amended at 55 FR 28883, July 16, 1990]

§ 3.41 - Handling.

(a) Any person who is subject to the Animal Welfare regulations and who moves live guinea pigs or hamsters from an animal holding area of a terminal facility to a primary conveyance or vice versa shall do so as quickly and efficiently as possible. Any person subject to the Animal Welfare Act and holding any live guinea pig or hamster in an animal holding area of a terminal facility or transporting any live guinea pig or hamster to or from a terminal facility shall provide the following:

(1) *Shelter from sunlight.* When sunlight is likely to cause overheating or discomfort, sufficient shade shall be provided to protect the live guinea pigs and hamsters from the direct rays of the sun and such live guinea pigs or hamsters shall not be subjected to surrounding air temperatures which exceed 29.5 °C. (85 °F.), and which shall be measured and read in the manner prescribed § 3.40 of this part, for a period of more than 45 minutes.

(2) *Shelter from rain or snow.* Live guinea pigs and hamsters shall be provided protection to allow them to remain dry during rain or snow.

(3) *Shelter from cold weather.* Transporting devices shall be covered to provide protection for live guinea pigs and hamsters when the outdoor air temperature falls below 10 °C. (50 °F.), and such live guinea pigs and hamsters shall not be subjected to surrounding air temperatures which fall

below 7.2 °C. (45 °F.), and which shall be measured and read in the manner prescribed in § 3.40 of this part, for a period of more than 45 minutes.

(b) Care shall be exercised to avoid handling of the primary enclosure in such a manner that may cause physical or emotional trauma to the live guinea pig or hamster contained therein.

(c) Primary enclosures used to transport any live guinea pig or hamster shall not be tossed, dropped, or needlessly tilted and shall not be stacked in a manner which may reasonably be expected to result in their falling.

[43 FR 21163, May 16, 1978, as amended at 43 FR 56216, Dec. 1, 1978; 55 FR 28883, July 16, 1990]

PART 3
Subpart B

Subpart C – Specifications for the Humane Handling, Care, Treatment and Transportation of Rabbits

FACILITIES AND OPERATING STANDARDS

§ 3.50 - Facilities, general.

(a) *Structural strength.* Indoor and outdoor housing facilities for rabbits shall be structurally sound and shall be maintained in good repair, to protect the animals from injury, to contain the animals, and to restrict the entrance of other animals.

(b) *Water and electric power.* Reliable and adequate electric power, if required to comply with other provisions of this subpart, and adequate potable water shall be available.

(c) *Storage.* Supplies of food and bedding shall be stored in facilities which adequately protect such supplies against infestation or contamination by vermin. Refrigeration shall be provided for supplies of perishable food.

(d) *Waste disposal.* Provision shall be made for the removal and disposal of animal and food wastes, bedding, dead animals, and debris. Disposal facilities shall be so provided and operated as to minimize vermin infestation, odors, and disease hazards.

(e) *Washroom and sinks.* Facilities, such as washrooms, basins, or sinks, shall be provided to maintain cleanliness among animal caretakers.

[32 FR 3273, Feb. 24, 1967, as amended at 44 FR 63492, Nov. 2, 1979]

PART 3
Subpart C

§ 3.51 - Facilities, indoor.

(a) *Heating.* Indoor housing facilities for rabbits need not be heated.

(b) *Ventilation.* Indoor housing facilities for rabbits shall be adequately ventilated to provide for the health and comfort of the animals at all times. Such facilities shall be provided with fresh air either by means of windows, doors, vents, or air conditioning and shall be ventilated so as to minimize drafts, odors, and moisture condensation. Auxiliary ventilation, such as exhaust fans and vents or air conditioning, shall be provided when the ambient temperature is 85 °F. or higher.

(c) *Lighting.* Indoor housing facilities for rabbits shall have ample light, by natural or artificial means, or both, of good quality and well distributed. Such lighting shall provide uniformly distributed illumination of sufficient light intensity to permit routine inspection and cleaning during the entire working period. Primary enclosures shall be so placed as to protect the rabbits from excessive illumination.

(d) *Interior surfaces.* The interior building surfaces of indoor housing facilities shall be constructed and maintained so that they are substantially impervious to moisture and may be readily sanitized.

§ 3.52 - Facilities, outdoor.

(a) *Shelter from sunlight.* When sunlight is likely to cause overheating or discomfort, sufficient shade shall be provided to allow all rabbits kept outdoors to protect themselves from the direct rays of the sun. When the atmospheric temperature exceeds 90 °F. artificial cooling shall be provided by a sprinkler system or other means.

(b) *Shelter from rain or snow.* Rabbits kept outdoors shall be provided with access to shelter to allow them to remain dry during rain or snow.

(c) *Shelter from cold weather.* Shelter shall be provided for all rabbits kept outdoors when the atmospheric temperature falls below 40 °F.

(d) *Protection from predators.* Outdoor housing facilities for rabbits shall be fenced or otherwise enclosed to minimize the entrance of predators.

(e) *Drainage.* A suitable method shall be provided to rapidly eliminate excess water.

§ 3.53 - Primary enclosures.

All primary enclosures for rabbits shall conform to the following requirements:

(a) *General.*

(1) Primary enclosures shall be structurally sound and maintained in good repair to protect the rabbits from injury, to contain them, and to keep predators out.

(2) Primary enclosures shall be constructed and maintained so as to enable the rabbits to remain dry and clean.

(3) Primary enclosures shall be constructed and maintained so that the rabbits contained therein have convenient access to clean food and water as required in this subpart.

(4) The floors of the primary enclosures shall be constructed so as to protect the rabbits' feet and legs from injury. Litter shall be provided in all primary enclosures having solid floors.

(5) A suitable nest box containing clean nesting material shall be provided in each primary enclosure housing a female with a litter less than one month of age.

(b) *Space requirements for primary enclosures acquired before August 15, 1990.* Primary enclosures shall be constructed and maintained so as to provide sufficient space for the animal to make normal postural adjustments with adequate freedom of movement. Each rabbit housed in a primary enclosure shall be provided a minimum amount of floor space, exclusive of the space taken up by food and water receptacles, in accordance with the following table:

Category	Individual weights (pounds)	Minimum space per rabbit (square inches)
Groups	3 through 5	144
	6 through 8	288
	9 or more	432
Individual adults	3 through 5	180
	6 through 8	360
	9 through 11	540
	12 or more	720
Nursing females	3 through 5	576
	6 through 8	720
	9 through 11	864
	12 or more	1080

(c) *Space requirements for primary enclosures acquired on or after August 15, 1990.*

(1) Primary enclosures shall be constructed and maintained so as to provide sufficient space for the animal to make normal postural adjustments with adequate freedom of movement.

(2) Each rabbit housed in a primary enclosure shall be provided a minimum amount of floor space, exclusive of the space taken up by food and water receptacles, in accordance with the following table:

	Individual weights		Minimum floor space		Minimum interior height	
	kg	lbs	m²	ft²	cm	in
Individual rabbits (weaned)	<2	<4.4	0.14	1.5	35.56	14
	2-4	4.4-8.8	0.28	3.0	35.56	14
	4-5.4	8.8-11.9	0.37	4.0	35.56	14
	>5.4	>11.9	0.46	5.0	35.56	14

	Weight of nursing female		Minimum floor space / female & litter		Minimum interior height	
	kg	lbs	m²	ft²	cm	in
Females with litters	<2	<4.4	0.37	4.0	35.56	14
	2-4	4.4-8.8	0.46	5.0	35.56	14
	4-5.4	8.8-11.9	0.56	6.0	35.56	14
	>5.4	>11.9	0.70	7.5	35.56	14

(3) Innovative primary enclosures that do not precisely meet the space requirements of paragraph (c)(2) of this section, but that do provide rabbits with a sufficient volume of space and the opportunity to express species-typical behavior, may be used at research facilities when approved by the Institutional Animal Care and Use Committee, and by dealers and exhibitors when approved by the Administrator.

[32 FR 3273, Feb. 24, 1967, as amended at 55 FR 28883, July 16, 1990]

ANIMAL HEALTH AND HUSBANDRY STANDARDS

§ 3.54 - Feeding.

(a) Rabbits shall be fed at least once each day except as otherwise might be required to provide adequate veterinary care. The food shall be free from contamination, wholesome, palatable and of sufficient quantity and nutritive value to meet the normal daily requirements for the condition and size of the rabbit.

(b) Food receptacles shall be accessible to all rabbits in a primary enclosure and shall be located so as to minimize contamination by excreta. All food receptacles shall be kept clean and sanitized at least once every 2 weeks. If self-feeders are used for the feeding of dry feed, measures must be taken to prevent molding, deterioration or caking of the feed.

PART 3
Subpart C

§ 3.55 - Watering.

Sufficient potable water shall be provided daily except as might otherwise be required to provide adequate veterinary care. All watering receptacles shall be sanitized when dirty: *Provided, however,* That such receptacles shall be sanitized at least once every 2 weeks.

§ 3.56 - Sanitation.

(a) *Cleaning of primary enclosures.*

(1) Primary enclosures shall be kept reasonably free of excreta, hair, cobwebs and other debris by periodic cleaning. Measures shall be taken to prevent the wetting of rabbits in such enclosures if a washing process is used.

(2) In primary enclosures equipped with solid floors, soiled litter shall be removed and replaced with clean litter at least once each week.

(3) If primary enclosures are equipped with wire or mesh floors, the troughs or pans under such enclosures shall be cleaned at least once each week. If worm bins are used under such enclosures they shall be maintained in a sanitary condition.

(b) *Sanitization of primary enclosures.*

(1) Primary enclosures for rabbits shall be sanitized at least once every 30 days in the manner provided in paragraph (b)(3) of this section.

(2) Prior to the introduction of rabbits into empty primary enclosures previously occupied, such enclosures shall be sanitized in the manner provided in paragraph (b)(3) of this section.

(3) Primary enclosures for rabbits shall be sanitized by washing them with hot water (180 °F.) and soap or detergent as in a mechanical cage washer, or by washing all soiled surfaces with a detergent solution followed by a safe and effective disinfectant, or by cleaning all soiled surfaces with live steam or flame.

(c) *Housekeeping.* Premises (buildings and grounds) shall be kept clean and in good repair in order to protect the animals from injury and to facilitate the prescribed husbandry practices set forth in this subpart. Premises shall remain free of accumulations of trash.

(d) *Pest control.* An effective program for the control of insects, ectoparasites, and avian and mammalian pests shall be established and maintained.

§ 3.57 - Employees.

A sufficient number of employees shall be utilized to maintain the prescribed level of husbandry practices set forth in this subpart. Such practices shall be under the supervision of an animal caretaker who has a background in animal husbandry or care.

§ 3.58 - Classification and separation.

Animals housed in the same primary enclosure shall be maintained in compatible groups, with the following additional restrictions:

(a) Rabbits shall not be housed in the same primary enclosure with any other species of animals unless required for scientific reasons.

(b) Rabbits under quarantine or treatment for a communicable disease shall be separated from other rabbits and other susceptible species of animals in such a manner as to minimize dissemination of such disease.

§ 3.59 - [Reserved]

TRANSPORTATION STANDARDS

Authority: Sections 3.60 through 3.66 issued under secs. 3, 5, 6, 10, 11, 14, 16, 17, 21; 80 Stat. 353; 84 Stat. 1561, 1562, 1563, 1564; 90 Stat. 418, 420, 423 (7 U.S.C. 2133, 2135, 2136, 2140, 2141, 2144, 2146, 2147, 2151); 37 FR 28464, 28477, 38 FR 19141.

PART 3

Subpart C

Source: Sections 3.60 through 3.66 appear at 42 FR 31565, June 21, 1977, unless otherwise noted.

§ 3.60 - Consignments to carriers and intermediate handlers.

(a) Carriers and intermediate handlers shall not accept any live rabbit presented by any dealer, research facility, exhibitor, operator of an auction sale, or other person, or any department, agency, or instrumentality of the United States or any State or local government for shipment, in commerce, more than 4 hours prior to the scheduled departure of the primary conveyance on which it is to be transported: *Provided, however,* That the carrier or intermediate handler and any dealer, research facility, exhibitor, operator of an auction sale, or other person, or any department, agency, or instrumentality of the United States or any State or local government may mutually agree to extend the time of acceptance to not more than 6 hours if specific prior scheduling of the animal shipment to destination has been made.

(b) Any carrier or intermediate handler shall only accept for transportation or transport, in commerce, any live rabbit in a primary enclosure which conforms to the requirements set forth in § 3.61 of the standards: *Provided, however,* That any carrier or intermediate handler may accept for transportation or transport, in commerce, any live rabbit consigned by any department, agency, or instrumentality of the United States having laboratory animal facilities or exhibiting animals or any licensed or registered dealer, research facility, exhibitor, or operator of any auction sale, if such consignor furnishes to the carrier or intermediate handler a certificate, signed by the consignor, stating that the primary enclosure complies with § 3.61 of the standards, unless such primary enclosure is obviously defective or damaged and it is apparent that it cannot reasonably be expected to contain the live rabbit without causing suffering or injury to such live rabbit. A copy of such certificate shall accompany the shipment to destination. The certificate shall include at least the following information:

(1) Name and address of the consignor;

(2) The number of rabbits in the primary enclosure(s);

(3) A certifying statement (e.g., "I hereby certify that the __ (number) primary enclosure(s) which are used to transport the animal(s) in this shipment complies (comply) with USDA standards for primary enclosures (9 CFR part 3)."); and

(4) The signature of the consignor, and date.

(c) Carriers or intermediate handlers whose facilities fail to meet the minimum temperature allowed by the standards may accept for transportation or transport, in commerce, any live rabbit consigned by any department, agency, or instrumentality of the United States or of any State or local government, or by any person (including any licensee or registrant under the Act, as well as any private individual) if the consignor furnishes to the carrier or intermediate handler a certificate executed by a veterinarian accredited by this Department pursuant to part 160 of this title on a specified date which shall not be more than 10 days prior to delivery of such rabbit for transportation in commerce, stating that such live rabbit is acclimated to air temperatures lower than those prescribed in §§ 3.65 and 3.66. A copy of such certificate shall accompany the shipment to destination. The certificate shall include at least the following information:

(1) Name and address of the consignor;

(2) The number of rabbits in the shipment;

(3) A certifying statement (e.g., "I hereby certify that the animal(s) in this shipment is (are), to the best of my knowledge, acclimated to air temperatures lower than 7.2 °C. (45 °F.).)"; and

(4) The signature of the USDA accredited veterinarian, assigned accreditation number, and date.

(d) Carriers and intermediate handlers shall attempt to notify the consignee at least once in every 6 hour period following the arrival of any live rabbit at the animal holding area of the terminal cargo facility. The time, date, and method of each attempted notification and the final notification to the consignee and the name of the person notifying the consignee shall be recorded on the copy of the shipping document retained by the carrier or intermediate handler and on a copy of the shipping document accompanying the animal shipment.

[42 FR 31565, June 21, 1977, as amended at 43 FR 21164, May 16, 1978; 44 FR 63493, Nov. 2, 1979]

PART 3
Subpart C

§ 3.61 - Primary enclosures used to transport live rabbits.

No person subject to the Animal Welfare regulations shall offer for transportation or transport in commerce any live rabbit in a primary enclosure that does not conform to the following requirements:

(a) Primary enclosures, such as compartments, transport cages, cartons, or crates, used to transport live rabbits shall be constructed in such a manner that:

(1) The structural strength of the enclosure shall be sufficient to contain the live rabbits and to withstand the normal rigors of transportation;

(2) The interior of the enclosure shall be free from any protrusions that could be injurious to the live rabbits contained therein;

(3) The openings of such enclosures are easily accessible at all times for emergency removal of the live rabbits;

(4) Except as provided in paragraph (h) of this section, there are ventilation openings located on two opposite walls of the primary enclosure

and the ventilation openings on each such wall shall be at least 16 percent of the total surface area of each such wall, or there are ventilation openings located on all four walls of the primary enclosure and the ventilation openings on each such wall shall be at least 8 percent of the total surface area of each such wall: *Provided, however,* That at least one-third of the total minimum area required for ventilation of the primary enclosure shall be located on the lower one-half of the primary enclosure and at least one-third of the total minimum area required for ventilation of the primary enclosure shall be located on the upper one-half of the primary enclosure;

(5) Except as provided in paragraph (h) of this section, projecting rims or other devices shall be on the exterior of the outside walls with any ventilation openings to prevent obstruction of the ventilation openings and to provide a minimum air circulation space 1.9 centimeters (.75 inch) between the primary enclosure and any adjacent cargo or conveyance wall; and

(6) Except as provided in paragraph (h) of this section, adequate handholds or other devices for lifting shall be provided on the exterior of the primary enclosure to enable the primary enclosure to be lifted without tilting and to ensure that the person handling the primary enclosure will not be in contact with the rabbit.

(b) Live rabbits transported in the same primary enclosure shall be maintained in compatible groups and shall not be transported in the same primary enclosure with other specie of animals.

(c) Primary enclosures used to transport live rabbits shall be large enough to ensure that each rabbit contained therein has sufficient space to turn about freely and to make normal postural adjustments.

(d) Not more than 15 live rabbits shall be transported in the same primary enclosure.

(e) Primary enclosures used to transport live rabbits as provided in this section shall have solid bottoms to prevent leakage in shipment and shall be cleaned and sanitized in a manner prescribed in § 3.56 of the standards, if previously used. Such primary enclosures shall contain clean litter of a suitable absorbent material which is safe and nontoxic to the rabbits, in sufficient quantity to absorb and cover excreta, unless the rabbits are on wire or other nonsolid floors.

(f) Primary enclosures used to transport live rabbits, except where such primary enclosures are permanently affixed in the animal cargo space of the primary conveyance, shall be clearly marked on top and on one or more sides with the works "Live Animal" in letters not less than 2.5 centimeters (1 inch) in height, and with arrows or other markings, to indicate the correct upright position of the container.

(g) Documents accompanying the shipment shall be attached in an easily accessible manner to the outside of a primary enclosure which is part of such shipment.

(h) When a primary enclosure is permanently affixed within the animal cargo space of the primary conveyance so that the front opening is the only source of ventilation for such primary enclosure, the front opening shall open directly to the outside or to an unobstructed aisle or passageway within the primary conveyance. Such front ventilation opening shall be at least 90 percent of the total surface area of the front wall of the primary enclosure and covered with bars, wire mesh or smooth expanded metal.

[42 FR 31565, June 21, 1977, as amended at 43 FR 21164, May 16, 1978; 55 FR 28883, July 16, 1990]

§ 3.62 - Primary conveyances (motor vehicle, rail, air, and marine).

(a) The animal cargo space of primary conveyances used in transporting live rabbits shall be designed and constructed to protect the health, and ensure the safety and comfort of the rabbits contained therein at all times.

(b) The animal cargo space shall be constructed and maintained in a manner to prevent the ingress of engine exhaust fumes and gases from the primary conveyance during transportation in commerce.

(c) No live rabbit shall be placed in an animal cargo space that does not have a supply of air sufficient for normal breathing for each live animal contained therein, and the primary enclosures shall be positioned in the animal cargo space in such a manner that each rabbit has access to sufficient air for normal breathing.

PART 3
Subpart C

(d) Primary enclosures shall be positioned in the primary conveyance in such a manner that in an emergency the live rabbits can be removed from the primary conveyance as soon as possible.

(e) The interior of the animal cargo space shall be kept clean.

(f) Live rabbits shall not be transported with any material, substance (e.g., dry ice) or device which may reasonably be expected to be injurious to the health and well-being of the rabbits unless proper precaution is taken to prevent such injury.

(g) The animal cargo space of primary conveyances used to transport rabbits shall be mechanically sound and provide fresh air by means of windows, doors, vents, or air conditioning so as to minimize drafts, odors, and moisture condensation. Auxiliary ventilation, such as fans, blowers, or air conditioners, shall be used in any cargo space containing live rabbits when the ambient temperature in the animal cargo space is 75 °F (23.9 °C) or higher. The ambient temperature within the animal cargo space shall not exceed 85 °F (29.5 °C) or fall below 45 °F (7.2 °C), except that the ambient temperature in the cargo space may be below 45 °F (7.2 °C) if the rabbits are accompanied by a certificate of acclimation to lower temperatures, as provided in § 3.60(c) of this part.

[42 FR 31565, June 21, 1977, as amended at 55 FR 28883, July 16, 1990]

§ 3.63 - Food and water requirements.

(a) If live rabbits are to be transported for a period of more than 6 hours, they shall have access to food and water or a type of food, which provides the requirements for food and water in quantity and quality sufficient to satisfy their food and water needs, during transit.

(b) Any dealer, research facility, exhibitor or operator of an auction sale offering any live rabbit to any carrier or intermediate handler for transportation, in commerce, shall provide an adequate supply of food or type of food, which provides the requirements for food and water, within the primary enclosure to meet the requirements of this section.

(c) No carrier or intermediate handler shall accept for transportation, in commerce, any live rabbit without an adequate supply of food or type of food, which provides the requirements for food and water, within the primary enclosure to meet the requirements of this section.

§ 3.64 - Care in transit.

(a) During surface transportation, it shall be the responsibility of the driver or other employee to visually observe the live rabbits as frequently as circumstances may dictate, but not less than once every 4 hours, to assure that they are receiving sufficient air for normal breathing, their ambient temperatures are within the prescribed limits, all other applicable standards are being complied with and to determine whether any of the live rabbits are in obvious physical distress and to provide any needed veterinary care as soon as possible. When transported by air, live rabbits shall be visually observed by the carrier as frequently as circumstances may dictate, but not less than once every 4 hours, if the cargo space is accessible during flight. If the animal cargo space is not accessible during flight, the carrier shall visually observe the live rabbits whenever loaded and unloaded and whenever the animal cargo space is otherwise accessible to assure that they are receiving sufficient air for normal breathing, their ambient temperatures are within the prescribed limits, all other applicable standards are being complied with and to determine whether any such live rabbits are in obvious physical distress. The carrier shall provide any needed veterinary care as soon as possible. No rabbit in obvious physical distress shall be transported in commerce.

(b) During the course of transportation, in commerce, live rabbits shall not be removed from their primary enclosures unless placed in other primary

162

enclosures or facilities conforming to the requirements provided in this subpart.

§ 3.65 - Terminal facilities.

No person subject to the Animal Welfare regulations shall commingle shipments of live rabbits with inanimate cargo. All animal holding areas of a terminal facility where shipments of rabbits are maintained shall be cleaned and sanitized as prescribed in § 3.56 of the standards often enough to prevent an accumulation of debris or excreta, to minimize vermin infestation, and to prevent a disease hazard. An effective program for the control of insects, ectoparasites, and avian and mammalian pests shall be established and maintained for all animal holding areas. Any animal holding area containing live rabbits shall be provided with fresh air by means of windows, doors, vents, or air conditioning and may be ventilated or air circulated by means of fans, blowers, or an air conditioning system so as to minimize drafts, odors, and moisture condensation. Auxiliary ventilation, such as exhaust fans and vents or fans or blowers or air conditioning shall be used for any animal holding area containing live rabbits when the air temperature within such animal holding area is 23.9 °C. (75 °F.) or higher. The air temperature around any live rabbit in any animal holding area shall not be allowed to fall below 7.2 °C. (45 °F.) nor be allowed to exceed 29.5 °C. (85 °F.) at any time. To ascertain compliance with the provisions of this paragraph, the air temperature around any live rabbit shall be measured and read outside the primary enclosure which contains such rabbit at a distance not to exceed .91 meters (3 feet) from any one of the external walls of the primary enclosure and on a level parallel to the bottom of such primary enclosure at a point which approximates half the distance between the top and bottom of such primary enclosure.

PART 3
Subpart C

[43 FR 56216, Dec. 1, 1978, as amended at 55 FR 28883, July 16, 1990]]

§ 3.66 - Handling.

(a) Any person who is subject to the Animal Welfare regulations and who moves live rabbits from an animal holding area of a terminal facility to a primary conveyance or vice versa shall do so as quickly and efficiently as possible. Any person subject to the Animal Welfare regulations and holding any live rabbit in an animal holding area of a terminal facility or transporting any live rabbit to or from a terminal facility shall provide the following:

(1) *Shelter from sunlight.* When sunlight is likely to cause overheating or discomfort, sufficient shade shall be provided to protect the live rabbits from the direct rays of the sun and such live rabbits shall not be subjected to surrounding air temperatures which exceed 29.5 °C. (85 °F.), and which shall be measured and read in the manner prescribed in § 3.65 of this part, for a period of more than 45 minutes.

(2) *Shelter from rain or snow.* Live rabbits shall be provided protection to allow them to remain dry during rain or snow.

(3*) Shelter from cold weather.* Transporting devices shall be covered to provide protection for live rabbits when the outdoor air temperature falls below 10 °C. (50 °F.), and such live rabbits shall not be subjected to surrounding air temperatures which fall below 7.2 °C. (45 °F.), and which shall be measured and read in the manner prescribed in § 3.65 of this part, for a period of more than 45 minutes unless such rabbits are accompanied by a certificate of acclimation to lower temperatures as prescribed in § 3.60(c).

(b) Care shall be exercised to avoid handling of the primary enclosure in such a manner that may cause physical or emotional trauma to the live rabbit contained therein.

(c) Primary enclosures used to transport any live rabbit shall not be tossed, dropped, or needlessly tilted and shall not be stacked in a manner which may reasonably be expected to result in their falling.

[43 FR 21164, May 16, 1978, as amended at 43 FR 56216, Dec. 1, 1978; 55 FR 28883, July 16, 1990]

PART 3

Subpart C

Subpart D – Specifications for the Humane Handling, Care, Treatment, and Transportation of Nonhuman Primates[2]

Source: 56 FR 6495, Feb. 15, 1991, unless otherwise noted.

FACILITIES AND OPERATING STANDARDS

§ 3.75 - Housing facilities, general.

(a) *Structure: construction.* Housing facilities for nonhuman primates must be designed and constructed so that they are structurally sound for the species of nonhuman primates housed in them. They must be kept in good repair, and they must protect the animals from injury, contain the animals securely, and restrict other animals from entering.

(b) *Condition and site.* Housing facilities and areas used for storing animal food or bedding must be free of any accumulation of trash, waste material, junk, weeds, and other discarded materials. Animal areas inside of housing facilities must be kept neat and free of clutter, including equipment, furniture, or stored material, but may contain materials actually used and necessary for cleaning the area, and fixtures and equipment necessary for proper husbandry practices and research needs. Housing facilities other than those maintained by research facilities and Federal research facilities must be physically separated from any other businesses. If a housing facility is located on the same premises as any other businesses, it must be physically separated from the other businesses so that animals the size of dogs, skunks, and raccoons, are prevented from entering it.

(c) *Surfaces.*

(1) *General requirements.* The surfaces of housing facilities – including perches, shelves, swings, boxes, houses, dens, and other furniture-type fixtures or objects within the facility – must be constructed in a manner and made of materials that allow them to be readily cleaned and sanitized, or removed or replaced when worn or soiled. Furniture-type fixtures or objects must be sturdily constructed and must be strong enough to provide for the safe activity and welfare of nonhuman primates. Floors may be made of dirt, absorbent bedding, sand, gravel, grass, or other similar material that can be readily cleaned, or can be removed or replaced whenever cleaning does not

PART 3
Subpart D

---

2 *Nonhuman primates include a great diversity of forms, ranging from the marmoset weighing only a few ounces, to the adult gorilla weighing hundreds of pounds, and include more than 240 species. They come from Asia, Africa, and Central and South America, and they live in different habitats in nature. Some have been transported to the United States from their natural habitats and some have been raised in captivity in the United States. Their nutritional and activity requirements differ, as do their social and environmental requirements. As a result, the conditions appropriate for one species do not necessarily apply to another. Accordingly, these minimum specifications must be applied in accordance with the customary and generally accepted professional and husbandry practices considered appropriate for each species, and necessary to promote their psychological well-being. These minimum standards apply only to live nonhuman primates, unless stated otherwise.*

eliminate odors, diseases, pests, insects, or vermin. Any surfaces that come in contact with nonhuman primates must:

(i) Be free of excessive rust that prevents the required cleaning and sanitization, or that affects the structural strength of the surface; and

(ii) Be free of jagged edges or sharp points that might injure the animals.

(2) *Maintenance and replacement of surfaces.* All surfaces must be maintained on a regular basis. Surfaces of housing facilities – including houses, dens, and other furniture-type fixtures and objects within the facility – that cannot be readily cleaned and sanitized, must be replaced when worn or soiled.

(3) *Cleaning.* Hard surfaces with which nonhuman primates come in contact must be spot-cleaned daily and sanitized in accordance with § 3.84 of this subpart to prevent accumulation of excreta or disease hazards. If the species scent mark, the surfaces must be sanitized or replaced at regular intervals as determined by the attending veterinarian in accordance with generally accepted professional and husbandry practices. Floors made of dirt, absorbent bedding, sand, gravel, grass, or other similar material, and planted enclosures must be raked or spot-cleaned with sufficient frequency to ensure all animals the freedom to avoid contact with excreta. Contaminated material must be removed or replaced whenever raking and spot cleaning does not eliminate odors, diseases, insects, pests, or vermin infestation. All other surfaces of housing facilities must be cleaned and sanitized when necessary to satisfy generally accepted husbandry standards and practices. Sanitization may be done by any of the methods provided in § 3.84(b)(3) of this subpart for primary enclosures.

(d) *Water and electric power.* The housing facility must have reliable electric power adequate for heating, cooling, ventilation, and lighting, and for carrying out other husbandry requirements in accordance with the regulations in this subpart. The housing facility must provide running potable water for the nonhuman primates' drinking needs. It must be adequate for cleaning and for carrying out other husbandry requirements.

PART 3
Subpart D

(e) *Storage.* Supplies of food and bedding must be stored in a manner that protects the supplies from spoilage, contamination, and vermin infestation. The supplies must be stored off the floor and away from the walls, to allow cleaning underneath and around the supplies. Food requiring refrigeration must be stored accordingly, and all food must be stored in a manner that prevents contamination and deterioration of its nutritive value. Only the food and bedding currently being used may be kept in animal areas, and when not in actual use, open food and bedding supplies must be kept in leakproof containers with tightly fitting lids to prevent spoilage and contamination. Substances that are toxic to the nonhuman primates but that are required for normal husbandry practices must not be stored in food storage and preparation areas, but may be stored in cabinets in the animal areas.

(f) *Drainage and waste disposal.* Housing facility operators must provide for regular and frequent collection, removal, and disposal of animal and food wastes, bedding, dead animals, debris, garbage, water, and any other fluids and wastes, in a manner that minimizes contamination and disease risk. Housing facilities must be equipped with disposal facilities and drainage systems that are constructed and operated so that animal wastes and water are rapidly eliminated and the animals stay dry. Disposal and drainage systems must minimize vermin and pest infestation, insects, odors, and disease hazards. All drains must be properly constructed, installed, and maintained. If closed drainage systems are used, they must be equipped with traps and prevent the backflow of gases and the backup of sewage onto the floor. If the facility uses sump ponds, settlement ponds, or other similar systems for drainage and animal waste disposal, the system must be located far enough away from the animal area of the housing facility to prevent odors, diseases, insects, pests, and vermin infestation. If drip or constant flow watering devices are used to provide water to the animals, excess water must be rapidly drained out of the animal areas by gutters or pipes so that the animals stay dry. Standing puddles of water in animal areas must be mopped up or drained so that the animals remain dry. Trash containers in housing facilities and in food storage and food preparation areas must be leakproof and must have tightly fitted lids on them at all times. Dead animals, animal parts, and animal waste must not be kept in food storage or food preparation areas, food freezers, food refrigerators, and animal areas.

(g) *Washrooms and sinks.* Washing facilities, such as washrooms, basins, sinks, or showers must be provided for animal caretakers and must be readily accessible.

§ 3.76 - Indoor housing facilities.

(a) *Heating, cooling, and temperature.* Indoor housing facilities must be sufficiently heated and cooled when necessary to protect nonhuman primates from temperature extremes and to provide for their health and well-being. The ambient temperature in the facility must not fall below 45 °F (7.2 °C) for more than 4 consecutive hours when nonhuman primates are present, and must not rise above 85 °F (29.5 °C) for more than 4 consecutive hours when nonhuman primates are present. The ambient temperature must be maintained at a level that ensures the health and well-being of the species housed, as directed by the attending veterinarian, in accordance with generally accepted professional and husbandry practices.

(b) *Ventilation.* Indoor housing facilities must be sufficiently ventilated at all times when nonhuman primates are present to provide for their health

PART 3
Subpart D

167

and well-being and to minimize odors, drafts, ammonia levels, and moisture condensation. Ventilation must be provided by windows, doors, vents, fans, or air conditioning. Auxiliary ventilation, such as fans, blowers, or air conditioning, must be provided when the ambient temperature is 85 °F (29.5 °C) or higher. The relative humidity maintained must be at a level that ensures the health and well-being of the animals housed, as directed by the attending veterinarian, in accordance with generally accepted professional and husbandry practices.

(c) *Lighting.* Indoor housing facilities must be lighted well enough to permit routine inspection and cleaning of the facility, and observation of the nonhuman primates. Animal areas must be provided a regular diurnal lighting cycle of either natural or artificial light. Lighting must be uniformly diffused throughout animal facilities and provide sufficient illumination to aid in maintaining good housekeeping practices, adequate cleaning, adequate inspection of animals, and for the well-being of the animals. Primary enclosures must be placed in the housing facility so as to protect the nonhuman primates from excessive light.

§ 3.77 - Sheltered housing facilities.

(a) *Heating, cooling, and temperature.* The sheltered part of sheltered housing facilities must be sufficiently heated and cooled when necessary to protect the nonhuman primates from temperature extremes, and to provide for their health and well-being. The ambient temperature in the sheltered part of the facility must not fall below 45 °F (7.2 °C) for more than 4 consecutive hours when nonhuman primates are present, and must not rise above 85 °F (29.5 °C) for more than 4 consecutive hours when nonhuman primates are present, unless temperatures above 85 °F (29.5 °C) are approved by the attending veterinarian, in accordance with generally accepted husbandry practices. The ambient temperature must be maintained at a level that ensures the health and well-being of the species housed, as directed by the attending veterinarian, in accordance with generally accepted professional and husbandry practices.

PART 3
Subpart D

(b) *Ventilation.* The sheltered part of sheltered animal facilities must be sufficiently ventilated at all times to provide for the health and well-being of nonhuman primates and to minimize odors, drafts, ammonia levels, and moisture condensation. Ventilation must be provided by windows, doors, vents, fans, or air conditioning. Auxiliary ventilation, such as fans, blowers, or air conditioning, must be provided when the ambient temperature is 85 °F (29.5 °C) or higher. The relative humidity maintained must be at a level that ensures the health and well-being of the species housed, as directed by the attending veterinarian, in accordance with generally accepted professional and husbandry practices.

(c) *Lighting.* The sheltered part of sheltered housing facilities must be lighted well enough to permit routine inspection and cleaning of the facility, and observation of the nonhuman primates. Animal areas must be provided a regular diurnal lighting cycle of either natural or artificial light. Lighting must be uniformly diffused throughout animal facilities and provide sufficient illumination to aid in maintaining good housekeeping practices, adequate cleaning, adequate inspection of animals, and for the well-being of the animals. Primary enclosures must be placed in the housing facility so as to protect the nonhuman primates from excessive light.

(d) *Shelter from the elements.* Sheltered housing facilities for nonhuman primates must provide adequate shelter from the elements at all times. They must provide protection from the sun, rain, snow, wind, and cold, and from any weather conditions that may occur.

(e) *Capacity: multiple shelters.* Both the sheltered part of sheltered housing facilities and any other necessary shelter from the elements must be sufficiently large to provide protection comfortably to each nonhuman primate housed in the facility. If aggressive or dominant animals are housed in the facility with other animals, there must be multiple shelters or other means to ensure that each nonhuman primate has access to shelter.

(f) *Perimeter fence.* On and after February 15, 1994, the outdoor area of a sheltered housing facility must be enclosed by a fence that is of sufficient height to keep unwanted species out. Fences less than 6 feet high must be approved by the Administrator. The fence must be constructed so that it protects nonhuman primates by restricting unauthorized humans, and animals the size of dogs, skunks, and raccoons from going through it or under it and having contact with the nonhuman primates. It must be of sufficient distance from the outside wall or fence of the primary enclosure to prevent physical contact between animals inside the enclosure and outside the perimeter fence. Such fences less than 3 feet in distance from the primary enclosure must be approved by the Administrator. A perimeter fence is not required if:

PART 3
Subpart D

(1) The outside walls of the primary enclosure are made of a sturdy, durable material such as concrete, wood, plastic, metal, or glass, and are high enough and constructed in a manner that restricts contact with or entry by humans and animals that are outside the sheltered housing facility; or

(2) The housing facility is surrounded by a natural barrier that restricts the nonhuman primates to the housing facility and protects them from contact with unauthorized humans and animals that are outside the sheltered housing facility, and the Administrator gives written permission

(g) *Public barriers.* Fixed public exhibits housing nonhuman primates, such as zoos, must have a barrier between the primary enclosure and the public at any time the public is present, that restricts physical contact between the public and the nonhuman primates. Nonhuman primates used in trained animal acts or in uncaged public exhibits must be under the direct control and supervision of an experienced handler or trainer at all times when the public is present. Trained nonhuman primates may be permitted physical contact with the public, as allowed under § 2.131, but only if they are under the direct control and supervision of an experienced handler or trainer at all times during the contact.

(Approved by the Office of Management and Budget under control number 0579-0093)

§ 3.78 - Outdoor housing facilities.

(a) *Acclimation.* Only nonhuman primates that are acclimated, as determined by the attending veterinarian, to the prevailing temperature and humidity at the outdoor housing facility during the time of year they are at the facility, and that can tolerate the range of temperatures and climatic conditions known to occur at the facility at that time of year without stress or discomfort, may be kept in outdoor facilities.

(b) *Shelter from the elements.* Outdoor housing facilities for nonhuman primates must provide adequate shelter from the elements at all times. It must provide protection from the sun, rain, snow, wind, and cold, and from any weather conditions that may occur. The shelter must safely provide heat to the nonhuman primates to prevent the ambient temperature from falling below 45 °F (7.2 °C), except as directed by the attending veterinarian and in accordance with generally accepted professional and husbandry practices.

(c) *Capacity: multiple shelters.* The shelter must be sufficiently large to comfortably provide protection for each nonhuman primate housed in the facility. If aggressive or dominant animals are housed in the facility with other animals there must be multiple shelters, or other means to ensure protection for each nonhuman primate housed in the facility.

PART 3
Subpart D

(d) *Perimeter fence.* On and after February 15, 1994, an outdoor housing facility must be enclosed by a fence that is of sufficient height to keep unwanted species out. Fences less than 6 feet high must be approved by the Administrator. The fence must be constructed so that it protects nonhuman primates by restricting unauthorized humans, and animals the size of dogs, skunks, and raccoons from going through it or under it and having contact with the nonhuman primates. It must be of sufficient distance from the outside wall or fence of the primary enclosure to prevent physical contact between animals inside the enclosure and outside the perimeter fence. Such fences less than 3 feet in distance from the primary enclosure must be approved by the Administrator. A perimeter fence is not required if:

(1) The outside walls of the primary enclosure are made of a sturdy, durable material such as concrete, wood, plastic, metal, or glass, and are high

170

enough and constructed in a manner that restricts contact with or entry by humans and animals that are outside the housing facility; or

(2) The housing facility is surrounded by a natural barrier that restricts the nonhuman primates to the housing facility and protects them from contact with unauthorized humans and animals that are outside the housing facility, and the Administrator gives written permission.

(e) *Public barriers.* Fixed public exhibits housing nonhuman primates, such as zoos, must have a barrier between the primary enclosure and the public at any time the public is present, in order to restrict physical contact between the public and the nonhuman primates. Nonhuman primates used in trained animal acts or in uncaged public exhibits must be under the direct control and supervision of an experienced handler or trainer at all times when the public is present. Trained nonhuman primates may be allowed physical contact with the public, but only if they are under the direct control and supervision of an experienced handler or trainer at all times during the contact.

(Approved by the Office of Management and Budget under control number 0579-0093)

§ 3.79 - Mobile or traveling housing facilities.

(a) *Heating, cooling, and temperature.* Mobile or traveling housing facilities must be sufficiently heated and cooled when necessary to protect nonhuman primates from temperature extremes and to provide for their health and well-being. The ambient temperature in the traveling housing facility must not fall below 45 °F (7.2 °C) for more than 4 consecutive hours when nonhuman primates are present, and must not rise above 85 °F (29.5 °C) for more than 4 consecutive hours when nonhuman primates are present. The ambient temperature must be maintained at a level that ensures the health and well-being of the species housed, as directed by the attending veterinarian, and in accordance with generally accepted professional and husbandry practices.

PART 3
Subpart D

(b) *Ventilation.* Traveling housing facilities must be sufficiently ventilated at all times when nonhuman primates are present to provide for the health and well-being of nonhuman primates and to minimize odors, drafts, ammonia levels, moisture condensation, and exhaust fumes. Ventilation must be provided by means of windows, doors, vents, fans, or air conditioning. Auxiliary ventilation, such as fans, blowers, or air conditioning, must be provided when the ambient temperature in the traveling housing facility is 85 °F (29.5 °C) or higher.

(c) *Lighting.* Mobile or traveling housing facilities must be lighted well enough to permit routine inspection and cleaning of the facility, and observation of the nonhuman primates. Animal areas must be provided a regular diurnal lighting cycle of either natural or artificial light. Lighting must be uniformly diffused throughout animal facilities and provide sufficient illumination to aid in maintaining good housekeeping practices, adequate cleaning, adequate inspection of animals, and for the well-being of the animals. Primary enclosures must be placed in the housing facility so as to protect the nonhuman primates from excessive light.

(d) *Public barriers.* There must be a barrier between a mobile or traveling housing facility and the public at any time the public is present, in order to restrict physical contact between the nonhuman primates and the public. Nonhuman primates used in traveling exhibits, trained animal acts, or in uncaged public exhibits must be under the direct control and supervision of an experienced handler or trainer at all times when the public is present. Trained nonhuman primates may be allowed physical contact with the public, but only if they are under the direct control and supervision of an experienced handler or trainer at all times during the contact.

§ 3.80 - Primary enclosures.

Primary enclosures for nonhuman primates must meet the following minimum requirements:

(a) *General requirements.*

(1) Primary enclosures must be designed and constructed of suitable materials so that they are structurally sound for the species of nonhuman primates contained in them. They must be kept in good repair.

(2) Primary enclosures must be constructed and maintained so that they:

(i) Have no sharp points or edges that could injure the nonhuman primates;

(ii) Protect the nonhuman primates from injury;

(iii) Contain the nonhuman primates securely and prevent accidental opening of the enclosure, including opening by the animal;

(iv) Keep other unwanted animals from entering the enclosure or having physical contact with the nonhuman primates;

(v) Enable the nonhuman primates to remain dry and clean;

(vi) Provide shelter and protection from extreme temperatures and weather conditions that may be uncomfortable or hazardous to the species of nonhuman primate contained;

(vii) Provide sufficient shade to shelter all the nonhuman primates housed in the primary enclosure at one time;

(viii) Provide the nonhuman primates with easy and convenient access to clean food and water;

PART 3

Subpart D

(ix) Enable all surfaces in contact with nonhuman primates to be readily cleaned and sanitized in accordance with § 3.84(b)(3) of this subpart, or replaced when worn or soiled;

(x) Have floors that are constructed in a manner that protects the nonhuman primates from injuring themselves; and

(xi) Provide sufficient space for the nonhuman primates to make normal postural adjustments with freedom of movement.

(b) *Minimum space requirements.* Primary enclosures must meet the minimum space requirements provided in this subpart.

These minimum space requirements must be met even if perches, ledges, swings, or other suspended fixtures are placed in the enclosure. Low perches and ledges that do not allow the space underneath them to be comfortably occupied by the animal will be counted as part of the floor space.

(1) Prior to February 15, 1994:

(i) Primary enclosures must be constructed and maintained so as to provide sufficient space to allow each nonhuman primate to make normal postural adjustments with adequate freedom of movement; and

(ii) Each nonhuman primate housed in a primary enclosure must be provided with a minimum floor space equal to an area at least three times the area occupied by the primate when standing on four feet.

(2) On and after February 15, 1994:

(i) The minimum space that must be provided to *each* nonhuman primate, whether housed individually or with other nonhuman primates, will be determined by the typical weight of animals of its species, except for brachiating species and great apes[3] and will be calculated by using the following table:[4]

---

[3] *The different species of nonhuman primates are divided into six weight groups for determining minimum space requirements, except that all brachiating species of any weight are grouped together since they require additional space to engage in species-typical behavior. The grouping provided is based upon the typical weight for various species and not on changes associated with obesity, aging, or pregnancy. These conditions will not be considered in determining a nonhuman primate's weight group unless the animal is obviously unable to make normal postural adjustments and movements within the primary enclosure. Different species of prosimians vary in weight and should be grouped with their appropriate weight group. They have not been included in the weight table since different species typically fall into different weight groups. Infants and juveniles of certain species are substantially lower in weight than adults of those species and require the minimum space requirements of lighter weight species, unless the animal is obviously unable to make normal postural adjustments and movements within the primary enclosure.*

[4] *Examples of the kinds of nonhuman primates typically included in each age group are:*
Group 1—marmosets, tamarins, and infants (less than 6 months of age) of various species.
Group 2—capuchins, squirrel monkeys and similar size species, and juveniles (6 months to 3 years of age) of various species.
Group 3—macaques and African species.
Group 4—male macaques and large African species.
Group 5—baboons and nonbrachiating species larger than 33.0 lbs. (15 kg.).
Group 6—great apes over 55.0 lbs. (25 kg.), except as provided in paragraph (b)(2)(ii) of this section, and brachiating species.

PART 3
Subpart D

Group	Weight		Floor area/animal		Height	
	lbs.	(kg.)	ft.²	(m²)	in.	(cm.)
1	Under 2.2	(under 1)	1.6	(0.15)	20	(50.8)
2	2.2-6.6	(1-3)	3.0	(0.28)	30	(76.2)
3	6.6-22.0	(3-10)	4.3	(0.40)	30	(76.2)
4	22.0-33.0	(10-15)	6.0	(0.56)	32	(81.28)
5	33.0-55.0	(15-25)	8.0	(0.74)	36	(91.44)
6	Over 55.0	(over 25)	25.1	(2.33)	84	(213.36)

(ii) Dealers. exhibitors, and research facilities, including Federal research facilities, must provide great apes weighing over 110 lbs. (50 kg) an additional volume of space in excess of that required for Group 6 animals as set forth in paragraph (b)(2)(i) of this section, to allow for normal postural adjustments.

(iii) In the case of research facilities, any exemption from these standards must be required by a research proposal or in the judgment of the attending veterinarian and must be approved by the Committee. In the case of dealers and exhibitors, any exemption from these standards must be required in the judgment of the attending veterinarian and approved by the Administrator.

(iv) When more than one nonhuman primate is housed in a primary enclosure, the minimum space requirement for the enclosure is the sum of the minimum floor area space required for each individual nonhuman primate in the table in paragraph (b)(2)(i) of this section, and the minimum height requirement for the largest nonhuman primate housed in the enclosure. Provided however, that mothers with infants less than 6 months of age may be maintained together in primary enclosures that meet the floor area space and height requirements of the mother.

PART 3
Subpart D

(c) Innovative primary enclosures not precisely meeting the floor area and height requirements provided in paragraphs (b)(1) and (b)(2) of this section, but that do provide nonhuman primates with a sufficient volume of space and the opportunity to express species-typical behavior, may be used at research facilities when approved by the Committee, and by dealers and exhibitors when approved by the Administrator.

(Approved by the Office of Management and Budget under control number 0579-0093)

§ 3.81 - Environment enhancement to promote psychological well-being.

Dealers, exhibitors, and research facilities must develop, document, and follow an appropriate plan for environment enhancement adequate to promote the psychological well-being of nonhuman primates. The plan must be in accordance with the currently accepted professional standards as cited in appropriate professional journals or reference guides, and as directed by the attending veterinarian. This plan must be made available to APHIS upon request, and, in the case of research facilities, to officials of any pertinent funding agency. The plan, at a minimum, must address each of the following:

(a) *Social grouping.* The environment enhancement plan must include specific provisions to address the social needs of nonhuman primates of species known to exist in social groups in nature. Such specific provisions must be in accordance with currently accepted professional standards, as cited in appropriate professional journals or reference guides, and as directed by the attending veterinarian. The plan may provide for the following exceptions:

(1) If a nonhuman primate exhibits vicious or overly aggressive behavior, or is debilitated as a result of age or other conditions (e.g., arthritis), it should be housed separately;

(2) Nonhuman primates that have or are suspected of having a contagious disease must be isolated from healthy animals in the colony as directed by the attending veterinarian. When an entire group or room of nonhuman primates is known to have or believed to be exposed to an infectious agent, the group may be kept intact during the process of diagnosis, treatment, and control.

(3) Nonhuman primates may not be housed with other species of primates or animals unless they are compatible, do not prevent access to food, water, or shelter by individual animals. and are not known to be hazardous to the health and well-being of each other. Compatibility of nonhuman primates must be determined in accordance with generally accepted professional practices and actual observations, as directed by the attending veterinarian, to ensure that the nonhuman primates are in fact compatible. Individually housed nonhuman primates must be able to see and hear nonhuman primates of their own or compatible species unless the attending veterinarian determines that it would endanger their health, safety, or well-being.

PART 3
Subpart D

(b) *Environmental enrichment.* The physical environment in the primary enclosures must be enriched by providing means of expressing noninjurious species-typical activities. Species differences should be considered when determining the type or methods of enrichment. Examples of environmental enrichments include providing perches, swings, mirrors, and other increased cage complexities; providing objects to manipulate; varied food items; using foraging or task-oriented feeding methods; and providing interaction with the care giver or other familiar and knowledgeable person consistent with personnel safety precautions.

(c) *Special considerations.* Certain nonhuman primates must be provided special attention regarding enhancement of their environment, based on the needs of the individual species and in accordance with the instructions of the attending veterinarian. Nonhuman primates requiring special attention are the following:

(1) Infants and young juveniles;

(2) Those that show signs of being in psychological distress through behavior or appearance;

(3) Those used in research for which the Committee-approved protocol requires restricted activity;

(4) Individually housed nonhuman primates that are unable to see and hear nonhuman primates of their own or compatible species; and

(5) Great apes weighing over 110 lbs. (50 kg). Dealers, exhibitors, and research facilities must include in the environment enhancement plan special provisions for great apes weighing over 110 lbs. (50 kg), including additional opportunities to express species-typical behavior.

(d) *Restraint devices.* Nonhuman primates must not be maintained in restraint devices unless required for health reasons as determined by the attending veterinarian or by a research proposal approved by the Committee at research facilities. Maintenance under such restraint must be for the shortest period possible. In instances where long-term (more than 12 hours) restraint is required, the nonhuman primate must be provided the opportunity daily for unrestrained activity for at least one continuous hour during the period of restraint, unless continuous restraint is required by the research proposal approved by the Committee at research facilities.

(e) *Exemptions.*

PART 3
Subpart D

(1) The attending veterinarian may exempt an individual nonhuman primate from participation in the environment enhancement plan because of its health or condition, or in consideration of its well-being. The basis of the exemption must be recorded by the attending veterinarian for each exempted nonhuman primate. Unless the basis for the exemption is a permanent condition, the exemption must be reviewed at least every 30 days by the attending veterinarian.

(2) For a research facility, the Committee may exempt an individual nonhuman primate from participation in some or all of the otherwise required environment enhancement plans for scientific reasons set forth in the research proposal. The basis of the exemption shall be documented in the approved proposal and must be reviewed at appropriate intervals as determined by the Committee, but not less than annually.

(3) Records of any exemptions must be maintained by the dealer, exhibitor, or research facility and must be made available to USDA officials or officials of any pertinent funding Federal agency upon request.

(Approved by the Office of Management and Budget under control number 0579-0093)

ANIMAL HEALTH AND HUSBANDRY STANDARDS

§ 3.82 - Feeding.

(a) The diet for nonhuman primates must be appropriate for the species, size, age, and condition of the animal, and for the conditions in which the nonhuman primate is maintained, according to generally accepted professional and husbandry practices and nutritional standards. The food must be clean, wholesome, and palatable to the animals. It must be of sufficient quantity and have sufficient nutritive value to maintain a healthful condition and weight range of the animal and to meet its normal daily nutritional requirements.

(b) Nonhuman primates must be fed at least once each day except as otherwise might be required to provide adequate veterinary care. Infant and juvenile nonhuman primates must be fed as often as necessary in accordance with generally accepted professional and husbandry practices and nutritional standards, based upon the animals' age and condition.

(c) Food and food receptacles, if used, must be readily accessible to all the nonhuman primates being fed. If members of dominant nonhuman primate or other species are fed together with other nonhuman primates, multiple feeding sites must be provided. The animals must be observed to determine that all receive a sufficient quantity of food.

(d) Food and food receptacles, if used, must be located so as to minimize any risk of contamination by excreta and pests. Food receptacles must be kept clean and must be sanitized in accordance with the procedures listed in § 3.84(b)(3) of this subpart at least once every 2 weeks. Used food receptacles must be sanitized before they can be used to provide food to a different nonhuman primate or social grouping of nonhuman primates. Measures must be taken to ensure there is no molding, deterioration, contamination, or caking or wetting of food placed in self-feeders.

PART 3
Subpart D

§ 3.83 - Watering.

Potable water must be provided in sufficient quantity to every nonhuman primate housed at the facility. If potable water is not continually available to the nonhuman primates, it must be offered to them as often as necessary to ensure their health and well-being, but no less than twice daily for at least 1 hour each time, unless otherwise required by the attending veterinarian, or as required by the research proposal approved by the Committee at research facilities. Water receptacles must be kept clean and sanitized in accordance with methods provided in § 3.84(b)(3) of this subpart at least once every 2 weeks or as often as necessary to keep them clean and free from contamination. Used water receptacles must be sanitized before they can be used to provide water to a different nonhuman primate or social grouping of nonhuman primates.

(Approved by the Office of Management and Budget under control number 0579-0093)

§ 3.84 - Cleaning, sanitization, housekeeping, and pest control.

(a) *Cleaning of primary enclosures.* Excreta and food waste must be removed from inside each indoor primary enclosure daily and from underneath them as often as necessary to prevent an excessive accumulation of feces and food waste, to prevent the nonhuman primates from becoming soiled, and to reduce disease hazards, insects, pests, and odors. Dirt floors, floors with absorbent bedding, and planted areas in primary enclosures must be spot-cleaned with sufficient frequency to ensure all animals the freedom to avoid contact with excreta, or as often as necessary to reduce disease hazards, insects, pests, and odors. When steam or water is used to clean the primary enclosure, whether by hosing, flushing, or other methods, nonhuman primates must be removed, unless the enclosure is large enough to ensure the animals will not be harmed, wetted, or distressed in the process. Perches, bars, and shelves must be kept clean and replaced when worn. If the species of the nonhuman primates housed in the primary enclosure engages in scent marking, hard surfaces in the primary enclosure must be spot-cleaned daily.

(b) *Sanitization of primary enclosures and food and water receptacles.*

PART 3
Subpart D

(1) A used primary enclosure must be sanitized in accordance with this section before it can be used to house another nonhuman primate or group of nonhuman primates.

(2) Indoor primary enclosures must be sanitized at least once every 2 weeks and as often as necessary to prevent an excessive accumulation of dirt, debris, waste, food waste, excreta, or disease hazard, using one of the methods prescribed in paragraph (b)(3) of this section. However, if the species of nonhuman primates housed in the primary enclosure engages in scent marking, the primary enclosure must be sanitized at regular intervals determined in accordance with generally accepted professional and husbandry practices.

(3) Hard surfaces of primary enclosures and food and water receptacles must be sanitized using one of the following methods:

(i) Live steam under pressure;

(ii) Washing with hot water (at least 180 °F (82.2 °C)) and soap or detergent, such as in a mechanical cage washer;

(iii) Washing all soiled surfaces with appropriate detergent solutions or disinfectants, or by using a combination detergent/disinfectant product that accomplishes the same purpose, with a thorough cleaning of the surfaces to remove organic material, so as to remove all organic material and mineral buildup, and to provide sanitization followed by a clean water rinse.

(4) Primary enclosures containing material that cannot be sanitized using the methods provided in paragraph (b)(3) of this section, such as sand, gravel, dirt, absorbent bedding, grass, or planted areas, must be sanitized by removing the contaminated material as necessary to prevent odors, diseases, pests, insects, and vermin infestation.

(c) *Housekeeping for premises.* Premises where housing facilities are located, including buildings and surrounding grounds, must be kept clean and in good repair in order to protect the nonhuman primates from injury, to facilitate the husbandry practices required in this subpart, and to reduce or eliminate breeding and living areas for rodents, pests, and vermin. Premises must be kept free of accumulations of trash, junk, waste, and discarded matter. Weeds, grass, and bushes must be controlled so as to facilitate cleaning of the premises and pest control.

(d) *Pest control.* An effective program for control of insects, external parasites affecting nonhuman primates, and birds and mammals that are pests, must be established and maintained so as to promote the health and well-being of the animals and reduce contamination by pests in animal areas.

§ 3.85 - Employees.

Every person subject to the Animal Welfare regulations (9 CFR parts 1, 2, and 3) maintaining nonhuman primates must have enough employees to carry out the level of husbandry practices and care required in this subpart. The employees who provide husbandry practices and care, or handle nonhuman primates, must be trained and supervised by an individual who has the knowledge, background, and experience in proper husbandry and care of nonhuman primates to supervise others. The employer must be certain that the supervisor can perform to these standards.

PART 3
Subpart D

TRANSPORTATION STANDARDS

§ 3.86 - Consignments to carriers and intermediate handlers.

(a) Carriers and intermediate handlers must not accept a nonhuman primate for transport in commerce more than 4 hours before the scheduled departure time of the primary conveyance on which the animal is to be transported. However, a carrier or intermediate handler may agree with anyone consigning a nonhuman primate to extend this time by up to 2 hours.

(b) Carriers and intermediate handlers must not accept a nonhuman primate for transport in commerce unless they are provided with the name, address, telephone number, and telex number, if applicable, of the consignee.

(c) Carriers and intermediate handlers must not accept a nonhuman primate for transport in commerce unless the consignor certifies in writing to the carrier or intermediate handler that the nonhuman primate was offered food and water during the 4 hours before delivery to the carrier or intermediate handler. The certification must be securely attached to the outside of the primary enclosure in a manner that makes it easily noticed and read. Instructions for no food or water are not acceptable unless directed by the attending veterinarian. Instructions must be in compliance with § 3.89 of this subpart. The certification must include the following information for each nonhuman primate:

(1) The consignor's name and address;

(2) The species of nonhuman primate;

(3) The time and date the animal was last fed and watered and the specific instructions for the next feeding(s) and watering(s) for a 24-hour period; and

(4) The consignor's signature and the date and time the certification was signed.

(d) Carriers and intermediate handlers must not accept a nonhuman primate for transport in commerce unless the primary enclosure meets the requirements of § 3.87 of this subpart. A carrier or intermediate handler must not accept a nonhuman primate for transport if the primary enclosure is obviously defective or damaged and cannot reasonably be expected to safely and comfortably contain the nonhuman primate without suffering or injury.

(e) Carriers and intermediate handlers must not accept a nonhuman primate for transport in commerce unless their animal holding area facilities meet the minimum temperature requirements provided in §§ 3.91 and 3.92 of this subpart, or unless the consignor provides them with a certificate signed by a veterinarian and dated no more than 10 days before delivery of the animal to the carrier or intermediate handler for transport in commerce, certifying that the animal is acclimated to temperatures lower than those that are required in §§ 3.91 and 3.92 of this subpart. Even if the carrier or intermediate handler receives this certification, the temperatures the nonhuman primate is exposed to while in the carrier's or intermediate handler's custody must not be lower than the minimum temperature specified by the veterinarian in accordance with paragraph (e)(4) of this section, and must be reasonably within the generally and professionally accepted temperature range for the nonhuman primate, as determined by the veterinarian, considering its age, condition, and species. A copy of the certification must accompany the nonhuman primate to its destination and must include the following information for each primary enclosure:

(1) The consignor's name and address;

(2) The number of nonhuman primates contained in the primary enclosure;

(3) The species of nonhuman primate contained in the primary enclosure;

(4) A statement by a veterinarian that to the best of his or her knowledge, each of the nonhuman primates contained in the primary enclosure is acclimated to air temperatures lower than 50 °F (10 °C), but not lower than a minimum temperature specified on the certificate based on the generally and professionally accepted temperature range for the nonhuman primate, considering its age, condition, and species; and

(5) The veterinarian's signature and the date the certification was signed.

(f) When a primary enclosure containing a nonhuman primate has arrived at the animal holding area of a terminal facility after transport, the carrier or intermediate handler must attempt to notify the consignee upon arrival and at least once in every 6-hour period after arrival. The time, date, and method of all attempted notifications and the actual notification of the consignee, and the name of the person who notifies or attempts to notify the consignee must be written either on the carrier's or intermediate handler's copy of the shipping document or on the copy that accompanies the primary enclosure. If the consignee cannot be notified within 24 hours after the nonhuman primate has arrived at the terminal facility, the carrier or intermediate handler must return the animal to the consignor or to whomever the consignor designates. If the consignee is notified of the arrival and does not take physical delivery of the nonhuman primate within 48 hours after arrival of the nonhuman primate, the carrier or intermediate handler must return the animal to the consignor or to whomever the consignor designates. The carrier or intermediate handler must continue to provide proper care, feeding, and housing to the nonhuman primate, and maintain the nonhuman primate in accordance with generally accepted professional and husbandry practices until the consignee accepts delivery of the nonhuman primate or until it is returned to the consignor or to whomever the consignor designates. The carrier or intermediate handler must obligate the consignor to reimburse the carrier or intermediate handler for the cost of return transportation and care.

PART 3
Subpart D

(Approved by the Office of Management and Budget under control number 0579-0093)

§ 3.87 - Primary enclosures used to transport nonhuman primates.

Any person subject to the Animal Welfare regulations (9 CFR parts 1, 2, and 3) must not transport or deliver for transport in commerce a nonhuman primate unless it is contained in a primary enclosure, such as a compartment, transport cage, carton, or crate, and the following requirements are met:

(a) *Construction of primary enclosures.* Primary enclosures used to transport nonhuman primates may be connected or attached to each other and must be constructed so that:

(1) The primary enclosure is strong enough to contain the nonhuman primate securely and comfortably and to withstand the normal rigors of transportation;

(2) The interior of the enclosure has no sharp points or edges and no protrusions that could injure the animal contained in it;

(3) The nonhuman primate is at all times securely contained within the enclosure and cannot put any part of its body outside the enclosure in a way that could result in injury to the animal, or to persons or animals nearby;

(4) The nonhuman primate can be easily and quickly removed from the enclosure in an emergency;

(5) The doors or other closures that provide access into the enclosure are secured with animal-proof devices that prevent accidental opening of the enclosure, including opening by the nonhuman primate;

(6) Unless the enclosure is permanently affixed to the conveyance, adequate devices such as handles or handholds are provided on its exterior, and enable the enclosure to be lifted without tilting it, and ensure that anyone handling the enclosure will not come into physical contact with the animal contained inside;

(7) Any material, treatment, paint, preservative, or other chemical used in or on the enclosure is nontoxic to the animal and not harmful to the health or well-being of the animal;

PART 3
Subpart D

(8) Proper ventilation is provided to the nonhuman primate in accordance with paragraph (c) of this section;

(9) Ventilation openings are covered with bars, wire mesh, or smooth expanded metal having air spaces; and

(10) The primary enclosure has a solid, leak-proof bottom, or a removable, leak-proof collection tray under a slatted or wire mesh floor that prevents seepage of waste products, such as excreta and body fluids, outside of the enclosure. If a slatted or wire mesh floor is used in the enclosure, it must be designed and constructed so that the animal cannot put any part of its body between the slats or through the holes in the mesh. It must contain enough previously unused litter to absorb and cover excreta. The litter must be of a suitably absorbent material that is safe and nontoxic to the nonhuman

primate and is appropriate for the species transported in the primary enclosure.

(b) *Cleaning of primary enclosures.* A primary enclosure used to hold or transport nonhuman primates in commerce must be cleaned and sanitized before each use in accordance with the methods provided in § 3.84(b)(3) of this subpart.

(c) *Ventilation.*

(1) If the primary enclosure is movable, ventilation openings must be constructed in one of the following ways:

(i) If ventilation openings are located on two opposite walls of the primary enclosure, the openings on each wall must be at least 16 percent of the total surface area of each such wall and be located above the midline of the enclosure; or

(ii) If ventilation openings are located on all four walls of the primary enclosure, the openings on every wall must be at least 8 percent of the total surface area of each such wall and be located above the midline of the enclosure.

(2) Unless the primary enclosure is permanently affixed to the conveyance, projecting rims or similar devices must be located on the exterior of each enclosure wall having a ventilation opening, in order to prevent obstruction of the openings. The projecting rims or similar devices must be large enough to provide a minimum air circulation space of 0.75 inches (1.9 centimeters) between the primary enclosure and anything the enclosure is placed against.

(3) If a primary enclosure is permanently affixed to the primary conveyance so that there is only a front ventilation opening for the enclosure, the primary enclosure must be affixed to the primary conveyance in such a way that the front ventilation opening cannot be blocked, and the front ventilation opening must open directly to an unobstructed aisle or passageway inside of the conveyance. The ventilation opening must be at least 90 percent of the total area of the front wall of the enclosure, and must be covered with bars, wire mesh, or smooth expanded metal having air spaces.

PART 3
Subpart D

(d) *Compatibility.*

(1) Only one live nonhuman primate may be transported in a primary enclosure, except as follows:

(i) A mother and her nursing infant may be transported together;

(ii) An established male-female pair or family group may be transported together, except that a female in estrus must not be transported with a male nonhuman primate;

(iii) A compatible pair of juveniles of the same species that have not reached puberty may be transported together.

(2) Nonhuman primates of different species must not be transported in adjacent or connecting primary enclosures.

(e) *Space requirements.* Primary enclosures used to transport nonhuman primates must be large enough so that each animal contained in the primary enclosure has enough space to turn around freely in a normal manner and to sit in an upright, hands down position without its head touching the top of the enclosure. However, certain larger species may be restricted in their movements, in accordance with professionally accepted standards of care, when greater freedom of movement would be dangerous to the animal, its handler, or to other persons.

(f) *Marking and labeling.* Primary enclosures, other than those that are permanently affixed to a conveyance, must be clearly marked in English on the top and on one or more sides with the words "Wild Animals," or "Live Animals," in letters at least 1 inch (2.5 cm.) high, and with arrows or other markings to indicate the correct upright position of the primary enclosure. Permanently affixed primary enclosures must be clearly marked in English with the words "Wild Animals" or "Live Animals," in the same manner.

(g) *Accompanying documents and records.* Shipping documents that must accompany shipments of nonhuman primates may be held by the operator of the primary conveyance, for surface transportation only, or must be securely attached in a readily accessible manner to the outside of any primary enclosure that is part of the shipment, in a manner that allows them to be detached for examination and securely reattached, such as in a pocket or sleeve. Instructions for administration of drugs, medication, and other special care must be attached to each primary enclosure in a manner that makes them easy to notice, to detach for examination, and to reattach securely. Food and water instructions must be attached in accordance with § 3.86(c) of this subpart.

PART 3

Subpart D

(Approved by the Office of Management and Budget under control number 0579-0093)

§ 3.88 - Primary conveyances (motor vehicle, rail, air, and marine).

(a) The animal cargo space of primary conveyances used to transport nonhuman primates must be designed, constructed, and maintained in a manner that at all times protects the health and well-being of the animals transported in it, ensures their safety and comfort, and prevents the entry of engine exhaust from the primary conveyance during transportation.

(b) The animal cargo space must have a supply of air that is sufficient for the normal breathing of all the animals being transported in it.

(c) Each primary enclosure containing nonhuman primates must be positioned in the animal cargo space in a manner that provides protection from the elements and that allows each nonhuman primate enough air for normal breathing.

(d) During air transportation, the ambient temperature inside a primary conveyance used to transport nonhuman primates must be maintained at a level that ensures the health and well-being of the species housed, in accordance with generally accepted professional and husbandry practices, at all times a nonhuman primate is present.

(e) During surface transportation, the ambient temperature inside a primary conveyance used to transport nonhuman primates must be maintained between 45 °F (7.2 °C) and 85 °F (30 °C) at all times a nonhuman primate is present.

(f) A primary enclosure containing a nonhuman primate must be placed far enough away from animals that are predators or natural enemies of nonhuman primates, whether the other animals are in primary enclosures or not, so that the nonhuman primate cannot touch or see the other animals.

(g) Primary enclosures must be positioned in the primary conveyance in a manner that allows the nonhuman primates to be quickly and easily removed from the primary conveyance in an emergency.

(h) The interior of the animal cargo space must be kept clean

(i) Nonhuman primates must not be transported with any material, substance (e.g., dry ice), or device in a manner that may reasonably be expected to harm the nonhuman primates or cause inhumane conditions.

§ 3.89 - Food and water requirements.

(a) Each nonhuman primate that is 1 year of age or more must be offered food[5] at least once every 24 hours. Each nonhuman primate that is less than 1 year of age must be offered food at least once every 12 hours. Each nonhuman primate must be offered potable water at least once every 12 hours. These time periods apply to dealers, exhibitors, and research facilities, including Federal research facilities, who transport nonhuman primates in their own primary conveyances, starting from the time the nonhuman primate was last offered food and potable water before transportation was begun. These time periods apply to carriers and intermediate handlers starting from the date and time stated on the certification provided under § 3.86(c) of this subpart. Each nonhuman primate must be offered food and potable water within 4 hours before being transported in commerce. Consignors who are subject to the Animal Welfare regulations (9 CFR parts 1, 2, and 3) must certify that each nonhuman primate was offered food and potable water within the 4 hours preceding delivery of the nonhuman primate to a carrier or intermediate handler for transportation in commerce, and must certify the

PART 3
Subpart D

---

5 _Proper food for purposes of this section is described in § 3.82 of this subpart, with the necessities and circumstances of the mode of travel taken into account._

date and time the food and potable water was offered, in accordance with § 3.86(c) of this subpart.

(b) Any dealer, exhibitor, or research facility, including a Federal research facility, offering a nonhuman primate to a carrier or intermediate handler for transportation in commerce must securely attach to the outside of the primary enclosure used for transporting the nonhuman primate, written instructions for a 24-hour period for the in-transit food and water requirements of the nonhuman primate(s) contained in the enclosure. The instructions must be attached in a manner that makes them easily noticed and read.

(c) Food and water receptacles must be securely attached inside the primary enclosure and placed so that the receptacles can be filled from outside of the enclosure without opening the door. Food and water receptacles must be designed, constructed, and installed so that a nonhuman primate cannot leave the primary enclosure through the food or water opening.

(Approved by the Office of Management and Budget under control number 0579-0093)

§ 3.90 - Care in transit.

(a) *Surface transportation (ground and water).* Any person subject to the Animal Welfare regulations (9 CFR parts 1, 2, and 3) transporting nonhuman primates in commerce must ensure that the operator of the conveyance or a person accompanying the operator of the conveyance observes the nonhuman primates as often as circumstances allow, but not less than once every 4 hours, to make sure that they have sufficient air for normal breathing, that the ambient temperature is within the limits provided in § 3.88(d) of this subpart, and that all other applicable standards of this subpart are being complied with. The regulated person transporting the nonhuman primates must ensure that the operator or the person accompanying the operator determines whether any of the nonhuman primates are in obvious physical distress, and obtains any veterinary care needed for the nonhuman primates at the closest available veterinary facility.

(b) *Air transportation.* During air transportation of nonhuman primates, it is the responsibility of the carrier to observe the nonhuman primates as frequently as circumstances allow, but not less than once every 4 hours if the animal cargo area is accessible during flight. If the animal cargo area is not accessible during flight, the carrier must observe the nonhuman primates whenever they are loaded and unloaded and whenever the animal cargo space is otherwise accessible to make sure that the nonhuman primates have sufficient air for normal breathing, that the ambient temperature is within the limits provided in § 3.88(d) of this subpart, and that all other applicable standards of this subpart are being complied with. The carrier must determine whether any of the nonhuman primates is in obvious physical distress, and

PART 3
Subpart D

arrange for any needed veterinary care for the nonhuman primates as soon as possible.

(c) If a nonhuman primate is obviously ill, injured, or in physical distress, it must not be transported in commerce, except to receive veterinary care for the condition.

(d) During transportation in commerce, a nonhuman primate must not be removed from its primary enclosure unless it is placed in another primary enclosure or a facility that meets the requirements of § 3.80 or § 3.87 of this subpart. Only persons who are experienced and authorized by the shipper, or authorized by the consignor or the consignee upon delivery, if the animal is consigned for transportation, may remove nonhuman primates from their primary enclosure during transportation in commerce, unless required for the health or well-being of the animal.

(e) The transportation regulations contained in this subpart must be complied with until a consignee takes physical delivery of the animal if the animal is consigned for transportation, or until the animal is returned to the consignor.

§ 3.91 - Terminal facilities.

(a) *Placement.* Any persons subject to the Animal Welfare regulations (9 CFR parts 1, 2, and 3) must not commingle shipments of nonhuman primates with inanimate cargo or with other animals in animal holding areas of terminal facilities. Nonhuman primates must not be placed near any other animals, including other species of nonhuman primates, and must not be able to touch or see any other animals, including other species of nonhuman primates.

(b) *Cleaning, sanitization, and pest control.* All animal holding areas of terminal facilities must be cleaned and sanitized in a manner prescribed in § 3.84(b)(3) of this subpart, as often as necessary to prevent an accumulation of debris or excreta and to minimize vermin infestation and disease hazards. Terminal facilities must follow an effective program in all animal holding areas for the control of insects, ectoparasites, and birds and mammals that are pests of nonhuman primates.

PART 3
Subpart D

(c) *Ventilation.* Ventilation must be provided in any animal holding area in a terminal facility containing nonhuman primates by means of windows, doors, vents, or air conditioning. The air must be circulated by fans, blowers, or air conditioning so as to minimize drafts, odors, and moisture condensation. Auxiliary ventilation, such as exhaust fans, vents, fans, blowers, or air conditioning, must be used in any animal holding area containing nonhuman primates when the ambient temperature is 85 °F (29.5 °C) or higher.

(d) *Temperature.* The ambient temperature in an animal holding area containing nonhuman primates must not fall below 45 °F (7.2 °C) or rise above 85 °F (29.5 °C) for more than four consecutive hours at any time nonhuman primates are present. The ambient temperature must be measured in the animal holding area by the carrier, intermediate handler, or a person transporting nonhuman primates who is subject to the Animal Welfare regulations (9 CFR parts 1, 2, and 3), outside any primary enclosure containing a nonhuman primate at a point not more than 3 feet (0.91 m.) away from an outside wall of the primary enclosure, on a level that is even with the enclosure and approximately midway up the side of the enclosure.

(e) *Shelter.* Any person subject to the Animal Welfare regulations (9 CFR parts 1, 2, and 3) holding a nonhuman primate in an animal holding area of a terminal facility must provide the following:

(1) **Shelter from sunlight and extreme heat.** Shade must be provided that is sufficient to protect the nonhuman primate from the direct rays of the sun.

(2) **Shelter from rain or snow.** Sufficient protection must be provided to allow nonhuman primates to remain dry during rain, snow, and other precipitation.

(f) *Duration.* The length of time any person subject to the Animal Welfare regulations (9 CFR parts 1, 2, and 3) can hold a nonhuman primate in an animal holding area of a terminal facility upon arrival is the same as that provided in § 3.86(f) of this subpart.

§ 3.92 - Handling.

(a) Any person subject to the Animal Welfare regulations (9 CFR parts 1, 2, and 3) who moves (including loading and unloading) nonhuman primates within, to, or from the animal holding area of a terminal facility or a primary conveyance must do so as quickly and efficiently as possible, and must provide the following during movement of the nonhuman primate:

(1) *Shelter from sunlight and extreme heat.* Sufficient shade must be provided to protect the nonhuman primate from the direct rays of the sun. A nonhuman primate must not be exposed to an ambient temperature above 85 °F (29.5 °C) for a period of more than 45 minutes while being moved to or from a primary conveyance or a terminal facility, The ambient temperature must be measured in the manner provided in § 3.91(d) of this subpart.

(2) *Shelter from rain or snow.* Sufficient protection must be provided to allow nonhuman primates to remain dry during rain, snow, and other precipitation.

(3) *Shelter from cold temperatures.* Transporting devices on which nonhuman primates are placed to move them must be covered to protect the animals when the outdoor temperature falls below 45 °F (7.2 °C). A

nonhuman primate must not be exposed to an ambient air temperature below 45 °F (7.2 °C) for a period of more than 45 minutes, unless it is accompanied by a certificate of acclimation to lower temperatures as provided in § 3.86(e) of this subpart. The ambient temperature must be measured in the manner provided in § 3.91(d) of this subpart.

(b) Any person handling a primary enclosure containing a nonhuman primate must use care and must avoid causing physical harm or distress to the nonhuman primate.

(1) A primary enclosure containing a nonhuman primate must not be placed on unattended conveyor belts or on elevated conveyor belts, such as baggage claim conveyor belts and inclined conveyor ramps that lead to baggage claim areas, at any time; except that a primary enclosure may be placed on inclined conveyor ramps used to load and unload aircraft if an attendant is present at each end of the conveyor belt.

(2) A primary enclosure containing a nonhuman primate must not be tossed, dropped, or needlessly tilted, and must not be stacked in a manner that may reasonably be expected to result in its falling. It must be handled and positioned in the manner that written instructions and arrows on the outside of the primary enclosure indicate.

(c) This section applies to movement of a nonhuman primate from primary conveyance to primary conveyance, within a primary conveyance or terminal facility, and to or from a terminal facility or a primary conveyance.

(Approved by the Office of Management and Budget under control number 0579-0093)

PART 3
Subpart D

Subpart E – Specifications for the Humane Handling, Care, Treatment, and Transportation of Marine Mammals

Source: 44 FR 36874, June 22, 1979, unless otherwise noted.

FACILITIES AND OPERATING STANDARDS

§ 3.100 - Special considerations regarding compliance and/or variance.

(a) All persons subject to the Animal Welfare Act who maintain or otherwise handle marine mammals in captivity must comply with the provisions of this subpart, except that they may apply for and be granted a variance,[6] by the Deputy Administrator, from one or more specified provisions of § 3.104. The provisions of this subpart shall not apply, however, in emergency circumstances where compliance with one or more requirements would not serve the best interest of the marine mammals concerned.

(b) An application for a variance must be made to the Deputy Administrator in writing. The request must include:

(1) The species and number of animals involved,

(2) A statement from the attending veterinarian concerning the age and health status of the animals involved, and concerning whether the granting of a variance would be detrimental to the marine mammals involved,

(3) Each provision of the regulations that is not met,

(4) The time period requested for a variance,

(5) The specific reasons why a variance is requested, and

(6) The estimated cost of coming into compliance, if construction is involved.

(c) After receipt of an application for a variance, the Deputy Administrator may require the submission in writing of a report by two experts recommended by the American Association of Zoological Parks and Aquariums and approved by the Deputy Administrator concerning potential adverse impacts on the animals involved or on other matters relating to the effects of the requested variance on the health and well-being of such marine mammals. Such a report will be required only in those cases when the Deputy Administrator determines that such expertise is necessary to determine whether the granting of a variance would cause a situation detrimental to the health and well-being of the marine mammals involved. The cost of such report is to be paid by the applicant.

(d) Variances granted for facilities because of ill or infirm marine mammals that cannot be moved without placing their well-being in jeopardy,

PART 3
Subpart E

---

6 *Written permission from the Deputy Administrator to operate as a licensee or registrant under the Act without being in full compliance with one or more specified provisions of § 3.104.*

or for facilities within 0.3048 meters (1 foot) of compliance with any space requirement may be granted for up to the life of the marine mammals involved. Otherwise, variances shall be granted for a period not exceeding July 30, 1986, *Provided, however,* That under circumstances deemed justified by the Deputy Administrator, a maximum extension of 1 year may be granted to attain full compliance. A written request for the extension must be received by the Deputy Administrator by May 30, 1986. Consideration for extension by the Deputy Administrator will be limited to unforeseen or unusual situations such as when necessary public funds cannot be allocated in an appropriate time frame for a facility to attain full compliance by July 30, 1986.

(e) The Deputy Administrator shall deny any application for a variance if he determines that it is not justified under the circumstances or that allowing it will be detrimental to the health and well-being of the marine mammals involved.

(f) Any facility housing marine mammals that does not meet all of the space requirements as of July 30, 1984, must meet all of the requirements by September 28, 1984, or may operate without meeting such requirements until action is taken on an application for a variance if the application is submitted to the Deputy Administrator on or before September 28, 1984.

(g) A research facility may be granted a variance from specified requirements of this subpart when such variance is necessary for research purposes and is fully explained in the experimental design. Any time limitation stated in this section shall not be applicable in such case.

[49 FR 26681, June 28, 1984; 63 FR 2, Jan. 2, 1998]

§ 3.101 - Facilities, general.
(a) *Construction requirements.*

(1) Indoor and outdoor housing facilities for marine mammals must be structurally sound and must be maintained in good repair to protect the animals from injury, to contain the animals within the facility, and to restrict the entrance of unwanted animals. Lagoon and similar natural seawater facilities must maintain effective barrier fences extending above the high tide water level, or other appropriate measures, on all sides of the enclosure not contained by dry land to fulfill the requirements of this section.

PART 3
Subpart E

(2) All marine mammals must be provided with protection from abuse and harassment by the viewing public by the use of a sufficient number of uniformed or readily identifiable employees or attendants to supervise the viewing public, or by physical barriers, such as fences, walls, glass partitions, or distance, or any combination of these.

(3) All surfaces in a primary enclosure must be constructed of durable, nontoxic materials that facilitate cleaning, and disinfection as appropriate, sufficient to maintain water quality parameters as designated in § 3.106. All surfaces must be maintained in good repair as part of a regular, ongoing maintenance program. All facilities must implement a written protocol on cleaning so that surfaces do not constitute a health hazard to animals.

(4) Facilities that utilize natural water areas, such as tidal basins, bays, or estuaries (subject to natural tidewater action), for housing marine mammals are exempt from the drainage requirements of paragraph (c)(1) of this section.

(b) *Water and power supply.* Reliable and adequate sources of water and electric power must be provided by the facility housing marine mammals. Written contingency plans must be submitted to and approved by the Deputy Administrator regarding emergency sources of water and electric power in the event of failure of the primary sources, when such failure could reasonably be expected to be detrimental to the good health and well-being of the marine mammals housed in the facility. Contingency plans must include, but not be limited to, specific animal evacuation plans in the event of a disaster and should describe back-up systems and/or arrangements for relocating marine mammals requiring artificially cooled or heated water. If the emergency contingency plan includes release of marine mammals, the plan must include provision for recall training and retrieval of such animals. Facilities handling marine mammals must also comply with the requirements of § 2.134 of this subchapter. *[Please see **Note** at the end of this section.]*

(c) *Drainage.*

(1) Adequate drainage must be provided for all primary enclosure pools and must be located so that all of the water

contained in such pools may be effectively eliminated when necessary for cleaning the pool or for other purposes. Drainage effluent from primary enclosure pools must be disposed of in a manner that complies with all applicable Federal, State, and local pollution control laws.

(2) Drainage must be provided for primary enclosures and areas immediately surrounding pools. All drain covers and strainers must be securely fastened in order to minimize the potential risk of animal entrapment. Drains must be located so as to rapidly eliminate excess water (except in pools). Drainage effluent must be disposed of in a manner that complies with all applicable Federal, State, and local pollution control laws.

PART 3
Subpart E

(d) *Storage.* Supplies of food must be stored in facilities that adequately protect such supplies from deterioration, spoilage (harmful microbial growth), and vermin or other contamination. Refrigerators and freezers (or chilled and/or iced coolers for under 12 hours) must be used for perishable food. No substances that are known to be or may be toxic or harmful to marine mammals may be stored or maintained in the marine mammal food storage or preparation areas, except that cleaning agents may be kept in secured cabinets designed and located to prevent food contamination. Food, supplements, and medications may not be used beyond commonly accepted shelf life or date listed on the label.

(e) *Waste disposal.* Provision must be made for the removal and disposal of animal and food wastes, dead animals, trash, and debris. Disposal facilities must be provided and operated in a manner that will minimize odors and the risk of vermin infestation and disease hazards. All waste disposal procedures must comply with all applicable Federal, State, and local laws pertaining to pollution control, protection of the environment, and public health.

(f) *Employee washroom facilities.* Washroom facilities containing basins, sinks, and, as appropriate, showers, must be provided and conveniently located to maintain cleanliness among employees, attendants, and volunteers. These facilities must be cleaned and sanitized daily.

(g) *Enclosure or pool environmental enhancements.* Any nonfood objects provided for the entertainment or stimulation of marine mammals must be of sufficient size and strength to not be ingestible, readily breakable, or likely to cause injury to marine mammals, and be able to be cleaned, sanitized, and/or replaced effectively.

[66 FR 251, Jan. 3, 2001, as amended at 77 FR 76824, Dec. 31, 2012]

Effective Date Note: At 78 FR 46255, July 31, 2013, §2.134 was stayed indefinitely, effective July 31, 2013.

§ 3.102 - Facilities, indoor.

(a) *Ambient temperature.* The air and water temperatures in indoor facilities shall be sufficiently regulated by heating or cooling to protect the marine mammals from extremes of temperature, to provide for their good health and well-being and to prevent discomfort, in accordance with the currently accepted practices as cited in appropriate professional journals or reference guides, depending upon the species housed therein. Rapid changes in air and water temperatures shall be avoided.

PART 3
Subpart E

(b) *Ventilation.* Indoor housing facilities shall be ventilated by natural or artificial means to provide a flow of fresh air for the marine mammals and to minimize the accumulation of chlorine fumes, other gases, and objectionable odors. A vertical air space averaging at least 1.83 meters (6 feet) shall be maintained in all primary enclosures housing marine mammals, including pools of water.

(c) *Lighting.* Indoor housing facilities for marine mammals shall have ample lighting, by natural or artificial means, or both, of a quality, distribution, and duration which is appropriate for the species involved. Sufficient lighting must be available to provide uniformly distributed illumination which is adequate to permit routine inspections, observations, and cleaning of all parts of the primary enclosure including any den areas. The lighting shall be designed so as to prevent overexposure of the marine mammals contained therein to excessive illumination.[7]

[44 FR 36874, June 22, 1979; 63 FR 2, Jan. 2, 1998]

§ 3.103 - Facilities, outdoor.

(a) *Environmental temperatures.* Marine mammals shall not be housed in outdoor facilities unless the air and water temperature ranges which they may encounter during the period they are so housed do not adversely affect their health and comfort. A marine mammal shall not be introduced to an outdoor housing facility until it is acclimated to the air and water temperature ranges which it will encounter therein. The following requirements shall be applicable to all outdoor pools.

(1) The water surface of pools in outdoor primary enclosures housing polar bears and ice or cold water dwelling species of pinnipeds shall be kept sufficiently free of solid ice to allow for entry and exit of the animals.

(2) The water surface of pools in outdoor primary enclosures housing cetaceans and sea otters shall be kept free of ice.

(3) No sirenian or warm water dwelling species of pinnipeds or cetaceans shall be housed in outdoor pools where water temperature cannot be maintained within the temperature range to meet their needs.

(b) *Shelter.* Natural or artificial shelter which is appropriate for the species concerned, when the local climatic conditions are taken into consideration, shall be provided for all marine mammals kept outdoors to afford them protection from the weather or from direct sunlight.

(c) *Perimeter fence.* On and after May 17, 2000, all outdoor housing facilities (i.e., facilities not entirely indoors) must be enclosed by a perimeter fence that is of sufficient height to keep animals and unauthorized persons out. Fences less than 8 feet high for polar bears or less than 6 feet high for other marine mammals must be approved in writing by the Administrator. The fence must be constructed so that it protects marine mammals by restricting animals and unauthorized persons from going through it or under it and having contact with the marine mammals, and so that it can function as a secondary containment system for the animals in the facility when appropriate. The fence must be of sufficient distance from the outside of the

PART 3
Subpart E

---

[7] *Lighting intensity and duration must be consistent with the general well-being and comfort of the animal involved. When possible, it should approximate the lighting conditions encountered by the animal in its natural environment. At no time shall the lighting be such that it will cause the animal discomfort or trauma.*

primary enclosure to prevent physical contact between animals inside the enclosure and animals or persons outside the perimeter fence. Such fences less than 3 feet in distance from the primary enclosure must be approved in writing by the Administrator. For natural seawater facilities, such as lagoons, the perimeter fence must prevent access by animals and unauthorized persons to the natural seawater facility from the abutting land, and must encompass the land portion of the facility from one end of the natural seawater facility shoreline as defined by low tide to the other end of the natural seawater facility shoreline defined by low tide. A perimeter fence is not required:

(1) Where the outside walls of the primary enclosure are made of sturdy, durable material, which may include certain types of concrete, wood, plastic, metal, or glass, and are high enough and constructed in a manner that restricts entry by animals and unauthorized persons and the Administrator gives written approval; or

(2) Where the outdoor housing facility is protected by an effective natural barrier that restricts the marine mammals to the facility and restricts entry by animals and unauthorized persons and the Administrator gives written approval; or

(3) Where appropriate alternative security measures are employed and the Administrator gives written approval; or

(4) For traveling facilities where appropriate alternative security measures are employed.

[44 FR 36874, June 22, 1979, as amended at 64 FR 56147, Oct. 18, 1999]

§ 3.104 - Space requirements.

(a) *General.* Marine mammals must be housed in primary enclosures that comply with the minimum space requirements prescribed by this part. These enclosures must be constructed and maintained so that the animals contained within are provided sufficient space, both horizontally and vertically, to be able to make normal postural and social adjustments with adequate freedom of movement, in or out of the water. (An exception to these requirements is provided in § 3.110(b) for isolation or separation for medical treatment and/or medical training.) Enclosures smaller than required by the standards may be temporarily used for nonmedical training, breeding, holding, and transfer purposes. If maintenance in such enclosures for nonmedical training, breeding, or holding is to last longer than 2 weeks, such extension must be justified in writing by the attending veterinarian on a weekly basis. If maintenance in such enclosures for transfer is to last longer than 1 week, such extension must be justified in writing by the attending veterinarian on a weekly basis. Any enclosure that does not meet the minimum space requirement for primary enclosures (including, but not limited to, medical

PART 3
Subpart E

pools or enclosures, holding pools or enclosures, and gated side pools smaller than the minimum space requirements) may not be used for permanent housing purposes. Rotating animals between enclosures that meet the minimum space requirements and enclosures that do not is not an acceptable means of complying with the minimum space requirements for primary enclosures.

(b) *Cetaceans.* Primary enclosures housing cetaceans shall contain a pool of water and may consist entirely of a pool of water. In determining the minimum space required in a pool holding cetaceans, four factors must be satisfied. These are MHD, depth, volume, and surface area. For the purposes of this subpart, cetaceans are divided into Group I cetaceans and Group II cetaceans as shown in Table III in this section.

(1)(i) *The required minimum horizontal dimension (MHD)* of a pool for Group I cetaceans shall be 7.32 meters (24.0 feet) or two times the average adult length of the longest species of Group I cetacean housed therein (as measured in a parallel or horizontal line, from the tip of its upper jaw, or from the most anterior portion of the head in bulbous headed animals, to the notch in the tail fluke[8]), whichever is greater; except that such MHD measurement may be reduced from the greater number by up to 20 percent if the amount of the reduction is added to the MHD at the 90-degree angle and if the minimum volume and surface area requirements are met based on an MHD of 7.32 meters (24.0 feet) or two times the average adult length of the longest species of Group I cetacean housed therein, whichever is greater.

(ii) The MHD of a pool for Group II cetaceans shall be 7.32 meters (24.0 feet) or four times the average adult length of the longest species of cetacean to be housed therein (as measured in a parallel or horizontal line from the tip of its upper jaw, or from the most anterior portion of the head in bulbous headed animals, to the notch in the tail fluke), whichever is greater; except that such MHD measurement may be reduced from the greater number by up to 20 percent if the amount of the reduction is added to the MHD at the 90-degree angle and if the minimum volume and surface area requirements are met based on an MHD of 7.32 meters (24.0 feet) or four times the average adult length of the longest species of Group II cetacean housed therein, whichever is greater.

(iii) In a pool housing a mixture of Group I and Group II cetaceans, the MHD shall be the largest required for any cetacean housed therein.

(iv) Once the required MHD has been satisfied, the pool size may be required to be adjusted to increase the surface area and volume when

PART 3
Subpart E

---

8 *The body length of a Monodon monoceros (narwhale) is measured from the tip of the upper incisor tooth to the notch in the tail fluke. If the upper incisor is absent or does not extend beyond the front of the head, then it is measured like other cetaceans, from the tip of the upper jaw to the notch in the tail fluke. Immature males should be anticipated to develop the "tusk" (usually left incisor tooth) beginning at sexual maturity.*

cetaceans are added. Examples of MHD and volume requirements for Group I cetaceans are shown in Table I, and for Group II cetaceans in Table II.

TABLE I – GROUP I CETACEANS[1]

Representative average adult lengths		Minimum horizontal dimension (MHD)		Minimum required depth		Volume of water required for each additional cetacean in excess of two	
Meters	Feet	Meters	Feet	Meters	Feet	Cubic meters	Cubic feet
1.68	5.5	7.32	24	1.83	6	8.11	284.95
2.29	7.5	7.32	24	1.83	6	15.07	529.87
2.74	9.0	7.32	24	1.83	6	21.57	763.02
3.05	10.0	7.32	24	1.83	6	26.73	942.00
3.51	11.5	7.32	24	1.83	6	35.40	1,245.79
3.66	12.0	7.32	24	1.83	6	38.49	1,356.48
4.27	14.0	8.53	28	2.13	7	60.97	2,154.04
5.49	18.0	10.97	36	2.74	9	129.65	4,578.12
5.64	18.5	11.28	37	2.82	9.25	140.83	4,970.33
5.79	19.0	11.58	38	2.90	9.50	152.64	5,384.32
6.71	22.0	13.41	44	3.36	11	237.50	8,358.68
6.86	22.5	13.72	45	3.43	11.25	253.42	8,941.64
7.32	24.0	14.63	48	3.66	12	307.89	10,851.84
8.53	28.0	17.07	56	4.27	14	487.78	17,232.32

1 All calculations are rounded off to the nearest hundredth. In converting the length of cetaceans from feet to meters, 1 foot equals .3048 meter. Due to rounding of meter figures as to the length of the cetacean, the correlation of meters to feet in subsequent calculations of MHD and additional volume of water required per cetacean, over two, may vary slightly from a strict feet to meters ratio. Cubic meters is based on: 1 cubic foot=0.0283 cubic meter.

TABLE II – GROUP II CETACEANS[1]

Representative average adult lengths		Minimum horizontal dimension (MHD)		Minimum required depth		Volume of water required for each additional cetacean in excess of four	
Meters	Feet	Meters	Feet	Meters	Feet	Cubic meters[1]	Cubic feet
1.52	5.0	7.32	24	1.83	6	13.28	471.00
1.68	5.5	7.32	24	1.83	6	16.22	569.91
1.83	6.0	7.32	24	1.83	6	19.24	678.24
2.13	7.0	8.53	28	1.83	6	26.07	923.16
2.29	7.5	9.14	30	1.83	6	30.13	1,059.75
2.44	8.0	9.75	32	1.83	6	34.21	1,205.76
2.59	8.5	10.36	34	1.83	6	38.55	1,361.19
2.74	9.0	10.97	36	1.83	6	43.14	1,526.04

1 Converting cubic feet to cubic meters is based on: 1 cubic foot=0.0283 of a cubic meter.

PART 3
Subpart E

TABLE III – AVERAGE ADULT LENGTHS OF MARINE MAMMALS MAINTAINED IN CAPTIVITY[1]

Species	Common name	Average adult length	
Group 1 Cetaceans:		**In meters**	**In feet**
Balaenoptera acutorostrata	Minke whale	8.50	27.9
Cephalorhynchus commersonii	Commerson's dolphin	1.52	5.0
Delphinapterus leucas	Beluga whale	4.27	14.0
Monodon monoceros	Narwhale	3.96	13.0
Globicephala melaena	Long-finned pilot whale	5.79	19.0
Globicephala macrorhynchus	Short-finned pilot whale	5.49	18.0
Grampus griseus	Risso's dolphin	3.66	12.0
Orcinus orca	Killer whale	7.32	24.0
Pseudorca carassidens	False killer whale	4.35	14.3
Tursiops truncatus (Atlantic)	Bottlenose dolphin	2.74	9.0
Tursiops truncatus (Pacific)	Bottlenose dolphin	3.05	10.0
Inia geoffrensis	Amazon porpoise	2.44	8.0
Phocoena phocoena	Harbor porpoise	1.68	5.5
Pontoporia blainvillei	Franciscana	1.52	5.0
Sotalia fluviatilis	Tucuxi	1.68	5.5
Platanista, all species	River dolphin	2.44	8.0

TABLE III – *continued*

PART 3
Subpart E

Group II Cetaceans:		In meters	In feet
Delphinus delphis	Common dolphin	2.59	8.5
Feresa attenuate	Pygmy killer whale	2.44	8.0
Kogia breviceps	Pygmy sperm whale	3.96	13.0
Kogia simus	Dwarf sperm whale	2.90	9.5
Lagenorhynchus acutus	Atlantic white-sided dolphin	2.90	9.5

Lagenorhynchus cruciger	Hourglass dolphin	1.70	5.6
Lagenorhynchus obliquidens	Pacific white-sided dolphin	2.29	7.5
Lagenorhynchus albirostris	White-beaked dolphin	2.74	9.0
Lagenorhynchus obscurus	Duskey dolphin	2.13	7.0
Lissodelphis borealis	Northern right whale dolphin	2.74	9.0
Neophocaena phocaenoides	Finless porpoise	1.83	6.0
Peponocephala electra	Melon-headed whale	2.74	9.0
Phocoenoides dalli	Dall's porpoise	2.00	6.5
Stenella longirostris	Spinner dolphin	2.13	7.0
Stenella coeruleoalba	Striped dolphin	2.29	7.5
Stenella attenuate	Spotted dolphin	2.29	7.5
Stenella plagiodon	Spotted dolphin	2.29	7.5
Steno bredanensis	Rough-toothed dolphin	2.44	8.0

1 This table contains the species of marine mammals known by the Department to be presently in captivity or that are likely to become captive in the future. Anyone who is subject to the Animal Welfare Act having species of marine mammals in captivity which are not included in this table should consult the Deputy Administrator with regard to the average adult length of such animals.

TABLE III – *continued*

Species	Common name	Average adult length			
		In meters		In feet	
Group 1 Pinnipeds:		**Male**	**Female**	**Male**	**Female**
Arctocephalus gazella**	Antarctic Fur Seal	1.80	1.20	5.9	3.9
Arctocephalus tropicalis**	Amsterdam Island Fur Seal	1.80	1.45	5.9	4.75
Arctocephalus australis**	South American Fur Seal	1.88	1.42	6.2	4.7
Arctocephalus pusillis**	Cape Fur Seal	2.73	1.83	8.96	6.0
Callorhinus ursinus**	Northern Fur Seal	2.20	1.45	7.2	4.75
Eumetopias jubatus**	Steller's Sea Lion	2.86	2.40	9.4	7.9

PART 3
Subpart E

Hydrurga leptonyx	Leopard Seal	2.90	3.30	9.5	10.8
*Mirounga angustirostris**	Northern Elephant Seal	3.96	2.49	13.0	8.2
*Mirounga leonina**	Southern Elephant Seal	4.67	2.50	15.3	8.2
*Odobenus rosmarus**	Walrus	3.15	2.60	10.3	8.5
*Otaria flavescens**	South American Sea Lion	2.40	2.00	7.9	6.6
Phoca caspica	Caspian Seal	1.45	1.40	4.75	4.6
Phoca fasciata	Ribbon Seal	1.75	1.68	5.7	5.5
Phoca larga	Harbor Seal	1.70	1.50	5.6	4.9
Phoca vitulina	Habor Seal	1.70	1.50	5.6	4.9
Zalophus californianus	California Sea Lion	2.24	1.75	7.3	5.7
*Halichoerus grypus**	Grar Seal	2.30	1.95	7.5	6.4
Phoca sibirica	Baikal Seal	1.70	1.85	5.6	6.1
Phoca groenlandica	Harp Seal	1.85	1.85	6.1	6.1
*Leptonychotes weddelli**	Weddell Seal	2.90	3.15	9.5	10.3
*Lobodon carcinophagus**	Crabeater Seal	2.21	2.21	7.3	7.3
*Ommatophoca rossi**	Ross Seal	1.99	2.13	6.5	7.0

TABLE III – *continued*

Group II Pinnipeds:		Male	Female	Male	Female
Erignathus barbatus	Bearded Seal	2.33	2.33	7.6	7.6
Phoca hispida	Ringed Seal	1.35	1.30	4.4	4.3
Cystophora cristata	Hooded Seal	2.60	2.00	8.5	6.6

*Note. **Any Group I animals maintained together will be considered as Group II when the animals maintained together include two or more sexually mature males from species marked with a double asterisk (**) regardless of whether the sexually mature males from the same species.*

TABLE III – *continued*

Species	Common name	Average adult length	
		In meters	In feet
Sirenia:			
Dugong dugong	Dugong	3.35	11.0
Trichechus manatus	West Indian Manatee	3.51	11.5
Trichechus inunguis	Amazon Manatee	2.44	8.0
Mustelidae:			
Enhydra lutris	Sea Otter	1.25	4.1

(2) *The minimum depth requirement* for primary enclosure pools for all cetaceans shall be one-half the average adult length of the longest species to be housed therein, regardless of Group I or Group II classification, or 1.83 meters (6.0 feet), whichever is greater, and can be expressed as d=L/2 or 6 feet, whichever is greater. Those parts of the primary enclosure pool which do not meet the minimum depth requirement cannot be included when calculating space requirements for cetaceans.

(3) *Pool volume.* A pool of water housing cetaceans which satisfies the MHD and which meets the minimum depth requirement, will have sufficient volume and surface area to hold up to two Group I cetaceans or up to four Group II cetaceans. If additional cetaceans are to be added to the pool, the volume as well as the surface area may have to be adjusted to allow for additional space necessary for such cetaceans. See Tables I, II, and IV for volumes and surface area requirements. The additional volume needed shall be based on the number and kind of cetaceans housed therein and shall be determined in the following manner.

(i) The minimum volume of water required for up to two Group I cetaceans is based upon the following formula:

$$\text{Volume} = \left(\frac{\text{MHD}}{2}\right)^2 \times 3.14 \times \text{depth}$$

PART 3
Subpart E

When there are more than two Group I cetaceans housed in a primary enclosure pool, the additional volume of water required for each additional Group I cetacean in excess of two is based on the following formula:

$$\text{Volume} = \left(\frac{\text{Average Adult Length}}{2}\right)^2 \times 3.14 \times \text{depth}$$

See Table I for required volumes.

203

(ii) The minimum volume of water required for up to four Group II cetaceans is based upon the following formula:

$$\text{Volume} = \left(\frac{\text{MHD}}{2}\right)^2 \times 3.14 \times \text{depth}$$

When there are more than four Group II cetaceans housed in a primary enclosure pool, the additional volume of water required for each additional Group II cetacean in excess of four is based on the following formula:

$$\text{Volume} = (\text{Avg. Adult Length})^2 \times 3.14 \times \text{depth}$$

See Table II for required volumes.

(iii) When a mixture of both Group I and Group II cetaceans are housed together, the MHD must be satisfied as stated in § 3.104(b)(1), and the minimum depth must be satisfied as stated in § 3.104(b)(2). Based on these figures, the resulting volume must then be calculated

$$\text{Volume} = \left(\frac{\text{MHD}}{2}\right)^2 \times 3.14 \times \text{depth}$$

Then the volume necessary for the cetaceans to be housed in the pool must be calculated (by obtaining the sum of the volumes required for each animal). If this volume is greater than that obtained by using the MHD and depth figures, then the additional volume required may be added by enlarging the pool in its lateral dimensions or by increasing its depth, or both. The minimum surface area requirements discussed next must also be satisfied.

(4)(i) *The minimum surface area requirements* for each cetacean housed in a pool, regardless of Group I or Group II classification, are calculated as follows:

$$\text{Surface Area} = \left(\frac{\text{Average Adult Body Length}}{2}\right)^2 \times 3.14 \times 1.5$$

$$\text{or: SA} = (\text{L}/2)2 \times 3.14 \times 1.5$$

PART 3

Subpart E

In a pool containing more than two Group I cetaceans or more than four Group II cetaceans,[9] the additional surface area which may be required when animals are added must be calculated for each such animal.

---

9 *A pool containing up to two Group I cetaceans or up to four Group II cetaceans which meets the required MHD and depth will have the necessary surface area and volume required for the animals contained therein.*

(ii) When a mixture of Group I and Group II cetaceans are to be housed in a pool, the required MHD, depth, and volume must be met. Then the required surface area must be determined for each animal in the pool. The sum of these surface areas must then be compared to the surface area which is obtained by a computation based on the required MHD of the pool.[10] The larger of the two figures represents the surface area which is required for a pool housing a mixture of Group I and Group II cetaceans. Pool surfaces where the depth does not meet the minimum requirements cannot be used in determining the required surface area.

(iii) Surface area requirements are given in Table IV.

TABLE IV – MINIMUM SURFACE AREA REQUIRED FOR EACH CETACEAN

Average adult length of each cetacean		Surface area required for each cetacean	
Meters	Feet	Sq. meters*	Sq. feet
1.68	5.5	3.31	33.62
2.13	7.0	5.36	57.70
2.29	7.5	6.15	66.23
2.59	8.5	7.90	85.07
2.74	9.0	8.86	95.38
3.05	10.0	10.94	117.75
3.51	11.5	14.47	155.72
3.66	12.0	15.75	169.56
4.27	14.0	21.44	230.79
5.49	18.0	35.44	381.51
5.64	18.5	37.43	403.00
5.79	19.0	39.49	425.08
6.71	22.0	52.94	569.91
6.86	22.5	55.38	596.11
7.32	24.0	63.01	678.24
8.53	28.0	85.76	923.16

* Square meter=square feet/9×0.8361.

PART 3
Subpart E

(c) Sirenians. Primary enclosures housing sirenians shall contain a pool of water and may consist entirely of a pool of water.

(1) The required MHD of a primary enclosure pool for sirenians shall be two times the average adult length of the longest species of sirenian to be

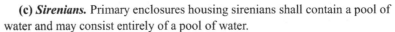

10 Since the MHD represents the diameter of a circle, the surface area based on the MHD is calculated by use of the following formula: $SA = \pi \times (MHD / 2)^2$.

housed therein. Calculations shall be based on the average adult length of such sirenians as measured in a horizontal line from the tip of the muzzle to the notch in the tail fluke of dugongs and from the tip of the muzzle to the most distal point in the rounded tail of the manatee.

(2) The minimum depth requirements for primary enclosure pools for all sirenians shall be one-half the average adult length of the longest species to be housed therein, or 1.52 meters (5.0 feet), whichever is greater. Those parts of the primary enclosure pool which do not meet the minimum depth requirements cannot be included when calculating space requirements for sirenians.

(3) A pool which satisfies the required MHD and depth shall be adequate for one or two sirenians. Volume and surface area requirements for additional animals shall be calculated using the same formula as for Group I cetaceans, except that the figure for depth requirement for sirenians shall be one-half the average adult length or 1.52 meters (5.0 feet), whichever is greater.

(d) *Pinnipeds.*

(1) Primary enclosures housing pinnipeds shall contain a pool of water and a dry resting or social activity area that must be close enough to the surface of the water to allow easy access for entering or leaving the pool. For the purposes of this subpart, pinnipeds have been divided into Group I pinnipeds and Group II pinnipeds as shown in Table III in this section. In certain instances some Group I pinnipeds shall be considered as Group II pinnipeds. (See Table III).

(2) The minimum size of the dry resting or social activity area of the primary enclosure for pinnipeds (exclusive of the pool of water) shall be based on the average adult length of each pinniped contained therein, as measured in a horizontal or extended position in a straight line from the tip of its nose to the tip of its tail. The minimum size of the dry resting or social activity area shall be computed using the following methods:

(i) *Group I pinnipeds.* Square the average adult length of each pinniped to be contained in the primary enclosure. Add the figures obtained for each of the pinnipeds in the primary enclosure to determine the dry resting or social activity area required for such pinnipeds. If only a single Group I pinniped is maintained in the primary enclosure, the minimum dry resting or social activity area shall be twice the square of the average adult length of that single Group I pinniped.

Examples:

(average adult length)^2 of 1st Group I pinniped+(average adult length)^2 of 2nd Group I pinniped=Total DRA for two pinnipeds

DRA for one pinniped=2×(average adult length of Group I pinniped)^2

(ii) *Group II pinnipeds.* List all pinnipeds contained in a primary enclosure by average adult length in descending order from the longest species of pinniped to the shortest species of pinniped. Square the average adult length of each pinniped. Multiply the average adult length squared of the longest pinniped by 1.5, the second longest by 1.4, the third longest by 1.3, the fourth longest by 1.2, and the fifth longest by 1.1, as indicated in the following example. Square the average adult length of the sixth pinniped and each additional pinniped. Add the figures obtained for all the pinnipeds in the primary enclosure to determine the required minimum dry resting or social activity area required for such pinnipeds. If only a single Group II pinniped is maintained in the primary enclosure, the minimum dry resting or social activity area must be computed for a minimum of two pinnipeds.

Examples: DRA for 1 Group II Pinniped = [(Average adult length)^2 × 1.5] + [(Average adult length)^2 × 1.4]

1st pinniped (avg. adult length)^2×1.5=social and DRA required
2nd pinniped (avg. adult length)^2×1.4=social and DRA required
3rd pinniped (avg. adult length)^1×1.3=social and DRA required
4th pinniped (avg. adult length)^2×1.2=social and DRA required
5th pinniped (avg. adult length)^2×1.1=social and DRA required
Each pinniped over 5 (avg. adult length)^2=social and DRA required

Total minimum social activity and dry resting area required
for all pinnipeds housed in a primary enclosure.

If all the pinnipeds in the primary enclosure are of the same species, the same descending order of calculation shall apply. Example: Hooded seal—average adult length of male=8.5 feet and female=6.6 feet. In a primary enclosure containing 2 males and 2 females, the social or DRA required would be the sum of [(8.5)^2 ×1.5] + [(8.5)^2 ×1.4] +[(6.6)^2 ×1.3] + [(6.6)^2 ×1.2].

If two or more sexually mature males are maintained together in a primary enclosure, the dry resting or social activity area shall be divided into two or more separate areas with sufficient visual barriers (such as fences, rocks, or foliage) to provide relief from aggressive animals.

(iii) *Mixture of Group I and Group II pinnipeds.* In a primary enclosure where a mixture of Group I and Group II pinnipeds is to be housed, the dry resting or social activity area shall be calculated as for Group II pinnipeds. The dry resting or social activity area shall be divided into two or

PART 3
Subpart E

more separate areas with sufficient visual barriers (such as fences, rocks, or foliage) to provide relief from aggressive animals.

(3)(i) The minimum surface area of a pool of water for pinnipeds shall be at least equal to the dry resting or social activity area required.

(ii) The MHD of the pool shall be at least one and one-half (1.5) times the average adult length of the largest species of pinniped to be housed in the enclosure; except that such MHD measurement may be reduced by up to 20 percent if the amount of the reduction is added to the MHD at the 90-degree angle.

(iii) The pool of water shall be at least 0.91 meters (3.0 feet) deep or one-half the average adult length of the longest species of pinniped contained therein, whichever is greater. Parts of the pool that do not meet the minimum depth requirement cannot be used in the calculation of the dry resting and social activity area, or as part of the MHD or required surface area of the pool.

(e) *Polar bears.* Primary enclosures housing polar bears shall consist of a pool of water, a dry resting and social activity area, and a den. A minimum of 37.16 square meters (400 square feet) of dry resting and social activity area shall be provided for up to two polar bears, with an additional 3.72 square meters (40 square feet) of dry resting and social activity area for each additional polar bear. The dry resting and social activity area shall be provided with enough shade to accommodate all of the polar bears housed in such primary enclosure at the same time. The pool of water shall have an MHD of not less than 2.44 meters (8.0 feet) and a surface area of at least 8.93 square meters (96.0 square feet) with a minimum depth of 1.52 meters (5.0 feet) with the exception of any entry and exit area. This size pool shall be adequate for two polar bears. For each additional bear, the surface area of the pool must be increased by 3.72 square meters (40 square feet). In measuring this additional surface area, parts of the pool which do not meet minimum depth cannot be considered. The den shall be at least 1.83 meters (6 feet) in width and depth and not less than 1.52 meters (5 feet) in height. It will be so positioned that the viewing public shall not be visible from the interior of the den. A separate den shall be provided for each adult female of breeding age which is permanently housed in the same primary enclosure with an adult male of breeding age. Female polar bears in traveling acts or shows must be provided a den when pregnancy has been determined.

(f) *Sea otters.*

(1) Primary enclosures for sea otters shall consist of a pool of water and a dry resting area. The MHD of the pool of water for sea otters shall be at least three times the average adult length of the sea otter contained therein (measured in a horizontal line from the tip of its nose to the tip of its tail) and the pool shall be not less than .91 meters (3.0 feet) deep. When more than two

PART 3
Subpart E

sea otters are housed in the same primary enclosure, additional dry resting area as well as pool volume is required to accommodate the additional sea otters. (See Table V).

(2) The minimum volume of water required for a primary enclosure pool for sea otters shall be based on the sea otter's average adult length. The minimum volume of water required in the pool shall be computed using the following method: Multiply the square of the sea otter's average adult length by 3.14 and then multiply the total by 0.91 meters (3.0 feet). This volume is satisfactory for one or two otters. To calculate the additional volume of water for each additional sea otter above two in a primary enclosure, multiply one-half of the square of the sea otter's average adult length by 3.14, then multiply by 0.91 meters (3.0 feet). (See Table V).

(3) The minimum dry resting area required for one or two sea otters shall be based on the sea otter's average adult length. The minimum dry resting area for one or two sea otters shall be computed using the following method: Square the average adult length of the sea otter and multiply the total by 3.14. When the enclosure is to contain more than two sea otters, the dry resting area for each additional animal shall be computed by multiplying one-half of the sea otter's average adult length by 3.14. Using 1.25 meters or 4.1 feet (the average adult length of a sea otter), the calculations for additional space will result in the following figures:

TABLE V – ADDITIONAL SPACE REQUIRED FOR EACH SEA OTTER WHEN MORE THAN TWO IN A PRIMARY ENCLOSURE

Average adult length of sea otter		Resting area		Pool Volume	
Meters	Feet	Square meters	Square Feet	Cubic meters	Cubic feet
1.25	4.1	1.96	6.44	2.23	79.17

[44 FR 36874, June 22, 1979, as amended at 45 FR 63261, Sept. 24, 1980; 49 FR 26682, 26685, June 28, 1984; 49 FR 27922, July 9, 1984; 63 FR 2, Jan. 2, 1998; 63 FR 47148, Sept. 4, 1998; 66 FR 252, Jan. 3, 2001]

PART 3
Subpart E

ANIMAL HEALTH AND HUSBANDRY STANDARDS

§ 3.105 - Feeding.

(a) The food for marine mammals must be wholesome, palatable, and free from contamination and must be of sufficient quantity and nutritive value to maintain marine mammals in a state of good health. The diet must be prepared with consideration for factors such as age, species, condition, and size of the marine mammal being fed. Marine mammals must be offered food at least once a day, except as directed by the attending veterinarian.

(b) Food receptacles, if used, must be located so as to be accessible to all marine mammals in the same primary enclosure and must be placed so as to minimize contamination of the food they contain. Such food receptacles must be cleaned and sanitized after each use.

(c) Food, when given to each marine mammal individually, must be given by an employee or attendant responsible to management who has the necessary knowledge to assure that each marine mammal receives an adequate quantity of food to maintain it in good health. Such employee or attendant is required to have the ability to recognize deviations from a normal state of good health in each marine mammal so that the food intake can be adjusted accordingly. Inappetence exceeding 24 hours must be reported immediately to the attending veterinarian. Public feeding may be permitted only in the presence and under the supervision of a sufficient number of knowledgeable, uniformed employees or attendants. Such employees or attendants must assure that the marine mammals are receiving the proper amount and type of food. Only food supplied by the facility where the marine mammals are kept may be fed to the marine mammals by the public. Marine mammal feeding records noting the estimated individual daily consumption must be maintained at the facility for a period of 1 year and must be made available for APHIS inspection. For marine mammals that are individually fed and not subject to public feeding, the feeding records should reflect an accurate account of food intake; for animals fed, in part, by the public, and for large, group-fed colonies of marine mammals where individual rations are not practical or feasible to maintain, the daily food consumption should be estimated as precisely as possible.

(d) Food preparation and handling must be conducted so as to assure the wholesomeness and nutritive value of the food. Frozen fish or other frozen food must be stored in freezers that are maintained at a maximum temperature of −18 °C (0 °F). The length of time food is stored and the method of storage, the thawing of frozen food, and the maintenance of thawed food must be conducted in a manner that will minimize contamination and that will assure that the food retains nutritive value and wholesome quality until the time of feeding. When food is thawed in standing or running water, cold water must be used. All foods must be fed to the marine mammals within 24 hours following the removal of such foods from the freezers for thawing, or if the food has been thawed under refrigeration, it must be fed to the marine mammals within 24 hours of thawing.

[66 FR 252, Jan. 3, 2001]

§ 3.106 - Water quality.

(a) *General.* The primary enclosure shall not contain water which would be detrimental to the health of the marine mammal contained therein.

(b) *Bacterial standards.*

(1) The coliform bacteria count of the primary enclosure pool shall not exceed 1,000 MPN (most probable number) per 100 ml. of water. Should a coliform bacterial count exceed 1,000 MPN, two subsequent samples may be taken at 48-hour intervals and averaged with the first sample. If such average count does not fall below 1,000 MPN, then the water in the pool shall be deemed unsatisfactory, and the condition must be corrected immediately.

(2) When the water is chemically treated, the chemicals shall be added so as not to cause harm or discomfort to the marine mammals.

(3) Water samples shall be taken and tested at least weekly for coliform count and at least daily for pH and any chemical additives (e.g. chlorine and copper) that are added to the water to maintain water quality standards. Facilities using natural seawater shall be exempt from pH and chemical testing unless chemicals are added to maintain water quality. However, they are required to test for coliforms. Records must be kept documenting the time when all such samples were taken and the results of the sampling. Records of all such test results shall be maintained by management for a 1-year period and must be made available for inspection purposes on request.

(c) *Salinity.* Primary enclosure pools of water shall be salinized for marine cetaceans as well as for those other marine mammals which require salinized water for their good health and well-being. The salinity of the water in such pools shall be maintained within a range of 15-36 parts per thousand.

(d) *Filtration and water flow.* Water quality must be maintained by filtration, chemical treatment, or other means so as to comply with the water quality standards specified in this section.

§ 3.107 - Sanitation.

(a) *Primary enclosures.*

(1) Animal and food waste in areas other than the pool of water must be removed from the primary enclosures at least daily, and more often when necessary, in order to provide a clean environment and minimize health and disease hazards.

(2) Particulate animal and food waste, trash, or debris that enters the primary enclosure pools of water must be removed at least daily, or as often as necessary, to maintain the required water quality and to minimize health and disease hazards to the marine mammals.

(3) The wall and bottom surfaces of the primary enclosure pools of water must be cleaned as often as necessary to maintain proper water quality. Natural organisms (such as algae, coelenterates, or molluscs, for

PART 3
Subpart E

example) that do not degrade water quality as defined in § 3.106, prevent proper maintenance, or pose a health or disease hazard to the animals are not considered contaminants.

(b) *Food preparation.* Equipment and utensils used in food preparation must be cleaned and sanitized after each use. Kitchens and other food handling areas where animal food is prepared must be cleaned at least once daily and sanitized at least once every week. Sanitizing must be accomplished by washing with hot water (8 °C, 180 °F, or higher) and soap or detergent in a mechanical dishwasher, or by washing all soiled surfaces with a detergent solution followed by a safe and effective disinfectant, or by cleaning all soiled surfaces with live steam. Substances such as cleansing and sanitizing agents, pesticides, and other potentially toxic agents must be stored in properly labeled containers in secured cabinets designed and located to prevent contamination of food storage preparation surfaces.

(c) *Housekeeping.* Buildings and grounds, as well as exhibit areas, must be kept clean and in good repair. Fences must be maintained in good repair. Primary enclosures housing marine mammals must not have any loose objects or sharp projections and/or edges which may cause injury or trauma to the marine mammals contained therein.

(d) *Pest control.* A safe and effective program for the control of insects, ectoparasites, and avian and mammalian pests must be established and maintained. Insecticides or other such chemical agents must not be applied in primary enclosures housing marine mammals except when deemed essential by an attending veterinarian.

[66 FR 253, Jan. 3, 2001]

§ 3.108 - Employees or attendants.

(a) A sufficient number of adequately trained employees or attendants, responsible to management and working in concert with the attending veterinarian, must be utilized to maintain the prescribed level of husbandry practices set forth in this subpart. Such practices must be conducted under the supervision of a marine mammal caretaker who has demonstrable experience in marine mammal husbandry and care.

(b) The facility will provide and document participation in and successful completion of a facility training course for such employees. This training course will include, but is not limited to, species appropriate husbandry techniques, animal handling techniques, and information on proper reporting protocols, such as recordkeeping and notification of veterinary staff for medical concerns.

(c) Any training of marine mammals must be done by or under the direct supervision of experienced trainers.

(d) Trainers and handlers must meet professionally recognized standards for experience and training.

[66 FR 253, Jan. 3, 2001]

§ 3.109 - Separation.

Marine mammals, whenever known to be primarily social in the wild, must be housed in their primary enclosure with at least one compatible animal of the same or biologically related species, except when the attending veterinarian, in consultation with the husbandry/training staff, determines that such housing is not in the best interest of the marine mammal's health or well-being. However, marine mammals that are not compatible must not be housed in the same enclosure. Marine mammals must not be housed near other animals that cause them unreasonable stress or discomfort or interfere with their good health. Animals housed separately must have a written plan, approved by the attending veterinarian, developed in consultation with the husbandry/training staff, that includes the justification for the length of time the animal will be kept separated or isolated, information on the type and frequency of enrichment and interaction, if appropriate, and provisions for periodic review of the plan by the attending veterinarian. Marine mammals that are separated for nonmedical purposes must be held in facilities that meet minimum space requirements as outlined in § 3.104.

[66 FR 253, Jan. 3, 2001]

§ 3.110 - Veterinary care.

(a) Newly acquired marine mammals must be isolated from resident marine mammals. Animals with a known medical history must be isolated unless or until the newly acquired animals can be reasonably determined to be in good health by the attending veterinarian. Animals without a known medical history must be isolated until it is determined that the newly acquired animals are determined to be in good health by the attending veterinarian. Any communicable disease condition in a newly acquired marine mammal must be remedied before it is placed with resident marine mammals, unless, in the judgment of the attending veterinarian, the potential benefits of a resident animal as a companion to the newly acquired animal outweigh the risks to the resident animal.

PART 3
Subpart E

(b) Holding facilities must be in place and available to meet the needs for isolation, separation, medical treatment, and medical training of marine mammals. Marine mammals that are isolated or separated for nonmedical purposes must be held in facilities that meet minimum space requirements as outlined in § 3.104. Holding facilities used only for medical treatment and medical training need not meet the minimum space requirements as outlined in § 3.104. Holding of a marine mammal in a medical treatment or medical training enclosure that does not meet minimum space requirements for periods longer than 2 weeks must be noted in the animal's medical record and the attending veterinarian must provide a justification in the animal's medical record. If holding in such enclosures for medical treatment and/ or medical training is to last longer than 2 weeks, such extension must be justified in writing by the attending veterinarian on a weekly basis. In natural lagoon or coastal enclosures where isolation cannot be accomplished, since water circulation cannot be controlled or isolated, separation of newly acquired marine mammals must be accomplished using separate enclosures situated within the facility to prevent direct contact and to minimize the risk of potential airborne and water cross-contamination between newly acquired and resident animals.

(c) Any holding facility used for medical purposes that has contained a marine mammal with an infectious or contagious disease must be cleaned and/or sanitized in a manner prescribed by the attending veterinarian. No healthy animals may be introduced into this holding facility prior to such cleaning and/or sanitizing procedures. Any marine mammal exposed to a contagious animal must be evaluated by the attending veterinarian and monitored and/or isolated for an appropriate period of time as determined by the attending veterinarian.

(d) Individual animal medical records must be kept and made available for APHIS inspection. These medical records must include at least the following information:

(1) Animal identification/name, a physical description, including any identifying markings, scars, etc., age, and sex; and

(2) Physical examination information, including but not limited to length, weight, physical examination results by body system, identification of all medical and physical problems with proposed plan of action, all diagnostic test results, and documentation of treatment.

(e) A copy of the individual animal medical record must accompany any marine mammal upon its transfer to another facility, including contract or satellite facilities.

(f) All marine mammals must be visually examined by the attending veterinarian at least semiannually and must be physically examined under the supervision of and when determined to be necessary by the attending veterinarian. All cetaceans and sirenians must be physically examined by the attending veterinarian at least annually, unless APHIS grants an exception from this requirement based on considerations related to the health and safety of the cetacean or sirenian. These examinations must include, but are not limited to, a hands-on physical examination, hematology and blood chemistry, and other diagnostic tests as determined by the attending veterinarian.

(g)(1) A complete necropsy, including histopathology samples, microbiological cultures, and other testing as appropriate, must be conducted by or under the supervision of the attending veterinarian on all marine mammals that die in captivity. A preliminary necropsy report must be prepared by the veterinarian listing all pathologic lesions observed. The final necropsy report must include all gross and histopathological findings, the results of all laboratory tests performed, and a pathological diagnosis.

(2) Necropsy records will be maintained at the marine mammal's home facility and at the facility at which it died, if different, for a period of 3 years and must be presented to APHIS inspectors when requested.

[66 FR 253, Jan. 3, 2001]

§ 3.111 - Swim-with-the-dolphin programs.

Swim-with-the-dolphin (SWTD) programs shall comply with the requirements in this section, as well as with all other applicable requirements of the regulations pertaining to marine mammals.

(a) *Space requirements.* The primary enclosure for SWTD cetaceans shall contain an interactive area, a buffer area, and a sanctuary area. None of these areas shall be made uninviting to the animals. Movement of cetaceans into the buffer or sanctuary area shall not be restricted in any way. Notwithstanding the space requirements set forth in § 3.104, each of the three areas required for SWTD programs shall meet the following space requirements:

(1) The horizontal dimension for each area must be at least three times the average adult body length of the species of cetacean used in the program;

(2) The minimum surface area required for each area is calculated as follows:

(i) *Up to two cetaceans:*

$$\text{Surface Area (SA)} = \left(\frac{3 \times \text{average adult body length (L)}}{2}\right)^2 \times 3.14$$

(ii) *Three cetaceans:*

$$SA = \left(\frac{3 \times L}{2}\right)^2 \times 3.14 \times 2$$

(iii) *Additional SA for each animal in excess of three*

$$SA = \left(\frac{2 \times L}{2}\right)^2 \times 3.14$$

PART 3
Subpart E

(3) The average depth for sea pens, lagoons, and similar natural enclosures at low tide shall be at least 9 feet. The average depth for any manmade enclosure or other structure not subject to tidal action shall be at least 9 feet. A portion of each area may be excluded when calculating the average depth, but the excluded portion may not be used in calculating whether the interactive, buffer, and sanctuary area meet the requirements of paragraphs (a)(1), (a)(2), and (a)(4) of this section.

(4) The minimum volume required for each animal is calculated as follows:

$$Volume = SA \times 9$$

(b) *Water clarity.* Sufficient water clarity shall be maintained so that attendants are able to observe cetaceans and humans at all times while within the interactive area. If water clarity does not allow these observations, the interactive sessions shall be canceled until the required clarity is provided.

(c) *Employees and attendants.* Each SWTD program shall have, at the minimum, the following personnel, with the following minimum backgrounds (each position shall be held by a separate individual, with a sufficient number of attendants to comply with § 3.111(e)(4)):

(1) Licensee or manager – at least one full-time staff member with at least 6 years experience in a professional or managerial position dealing with captive cetaceans;

(2) Head trainer/behaviorist – at least one full-time staff member with at least 6 years experience in training cetaceans for SWTD behaviors in the past 10 years, or an equivalent amount of experience involving in-water training of cetaceans, who serves as the head trainer for the SWTD program;

(3) Trainer/supervising attendant – at least one full-time staff member with at least 3 years training and/or handling experience involving human/cetacean interaction programs;

(4) Attendant – an adequate number of staff members who are adequately trained in the care, behavior, and training of the program animals. Attendants shall be designated by the trainer, in consultation with the head trainer/behaviorist and licensee/manager, to conduct and monitor interactive sessions in accordance with § 3.111(e); and

(5) Attending veterinarian – at least one staff or consultant veterinarian who has at least the equivalent of 2 years full-time experience (4,160 or more hours) with cetacean medicine within the past 10 years, and who is licensed to practice veterinary medicine.

(d) *Program animals.* Only cetaceans that meet the requirements of § 3.111(e)(2) and (3) may be used in SWTD programs.

(e) *Handling.*

(1) Interaction time (i.e., designated interactive swim sessions) for each cetacean shall not exceed 2 hours per day. Each program cetacean shall have at least one period in each 24 hours of at least 10 continuous hours without public interaction.

(2) All cetaceans used in an interactive session shall be adequately trained and conditioned in human interaction so that they respond in the session to the attendants with appropriate behavior for safe interaction. The head trainer/behaviorist, trainer/supervising attendant, or attendant shall, at all times, control the nature and extent of the cetacean interaction with the public during a session, using the trained responses of the program animal.

(3) All cetaceans used in interactive sessions shall be in good health, including, but not limited to, not being infectious. Cetaceans undergoing veterinary treatment may be used in interactive sessions only with the approval of the attending veterinarian.

(4) The ratio of human participants to cetaceans shall not exceed 3:1. The ratio of human participants to attendants or other authorized SWTD personnel (i.e., head trainer/behaviorist or trainer/supervising attendant) shall not exceed 3:1.

(5) Prior to participating in an SWTD interactive session, members of the public shall be provided with oral and written rules and instructions for the session, to include the telephone and FAX numbers for APHIS, Animal Care, for reporting injuries or complaints. Members of the public shall agree, in writing, to abide by the rules and instructions before being allowed to participate in the session. Any participant who fails to follow the rules or instructions shall be removed from the session by the facility.

(6) All interactive sessions shall have at least two attendants or other authorized SWTD personnel (i.e., head trainer/behaviorist or trainer/supervising attendant). At least one attendant shall be positioned out of the water. One or more attendants or other authorized SWTD personnel may be positioned in the water. If a facility has more than two incidents during interactive sessions within a year's time span that have been dangerous or harmful to either a cetacean or a human, APHIS, in consultation with the head trainer/behaviorist, will determine if changes in attendant positions are needed.

(7) All SWTD programs shall limit interaction between cetaceans and humans so that the interaction does not harm the cetaceans, does not remove the element of choice from the cetaceans by actions such as, but not limited to, recalling the animal from the sanctuary area, and does not elicit unsatisfactory, undesirable, or unsafe behaviors from the cetaceans. All SWTD programs shall prohibit grasping or holding of the cetacean's body, unless under the direct and explicit instruction of an attendant eliciting a

PART 3
Subpart E

217

specific cetacean behavior, and shall prevent the chasing or other harassment of the cetaceans.

(8) In cases where cetaceans used in an interactive session exhibit unsatisfactory, undesirable, or unsafe behaviors, including, but not limited to, charging, biting, mouthing, or sexual contact with humans, such cetaceans shall either be removed from the interactive area or the session shall be terminated. Written criteria shall be developed by each SWTD program, and shall be submitted to and approved by APHIS[11] regarding conditions and procedures for maintaining compliance with paragraph (e)(4) of this section; for the termination of a session when removal of a cetacean is not possible; and regarding criteria and protocols for handling program animal(s) exhibiting unsatisfactory, undesirable, or unsafe behaviors, including retraining time and techniques, and removal from the program and/or facility, if appropriate. The head trainer/behaviorist shall determine when operations will be terminated, and when they may resume. In the absence of the head trainer/behaviorist, the determination to terminate a session shall be made by the trainer/supervising attendant. Only the head trainer/behaviorist may determine when a session may be resumed.

(f) *Recordkeeping.*

(1) Each facility shall provide APHIS[12] 12 with a description of its program at least 30 days prior to initiation of the program, or in the case of any program in place before September 4, 1998, not later than October 5, 1998. The description shall include at least the following:

(i) Identification of each cetacean in the program, by means of name and/or number, sex, age, and any other means the Administrator determines to be necessary to adequately identify the cetacean;

(ii) A description of the educational content and agenda of planned interactive sessions, and the anticipated average and maximum frequency and duration of encounters per cetacean per day;

(iii) The content and method of pre-encounter orientation, rules, and instructions, including restrictions on types of physical contact with the cetaceans;

(iv) A description of the SWTD facility, including the primary enclosure and other SWTD animal housing or holding enclosures at the facility;

(v) A description of the training, including actual or expected number of hours each cetacean has undergone or will undergo prior to participation in the program;

PART 3
Subpart E

---

11 Send to Administrator, c/o Animal and Plant Health Inspection Service, Animal Care, 4700 River Road Unit 84, Riverdale, Maryland 20737-1234.

12 See footnote 1 in § 3.111(e)(8).

(vi) The resume of the licensee and/or manager, the head trainer/ behaviorist, the trainer/supervising attendant, any other attendants, and the attending veterinarian;

(vii) The current behavior patterns and health of each cetacean, to be assessed and submitted by the attending veterinarian;

(viii) For facilities that employ a part-time attending veterinarian or consultant arrangements, a written program of veterinary care (APHIS form 7002), including protocols and schedules of professional visits; and

(ix) A detailed description of the monitoring program to be used to detect and identify changes in the behavior and health of the cetaceans.

(2) All SWTD programs shall comply in all respects with the regulations and standards set forth in parts 2 and 3 of this subchapter.

(3) Individual animal veterinary records, including all examinations, laboratory reports, treatments, and necropsy reports shall be kept at the SWTD site for at least 3 years and shall be made available to an APHIS official upon request during inspection.

(4) The following records shall be kept at the SWTD site for at least 3 years and shall be made available to an APHIS official upon request during inspection:

(i) Individual cetacean feeding records; and

(ii) Individual cetacean behavioral records.

(5) The following reports shall be kept at the SWTD site for at least 3 years and shall be made available to an APHIS official upon request during inspection:

(i) Statistical summaries of the number of minutes per day that each animal participated in an interactive session;

(ii) A statistical summary of the number of human participants per month in the SWTD program; and

(6) A description of any changes made in the SWTD program, which shall be submitted to APHIS[13] on a semi-annual basis.

(7) All incidents resulting in injury to either cetaceans or humans participating in an interactive session, which shall be reported to APHIS within 24 hours of the incident.[14] Within 7 days of any such incident, a written report shall be submitted to the Administrator.[15] The report shall provide a detailed description of the incident and shall establish a plan of action for the prevention of further occurrences.

PART 3
Subpart E

(g) *Veterinary care.*

(1) The attending veterinarian shall conduct on-site evaluations of each cetacean at least once a month. The evaluation shall include a visual inspection of the animal; examination of the behavioral, feeding, and medical

---

13 See footnote 1 in § 3.111(e)(8).

14 Eastern Regional Office: (919) 855-7100. Western Regional Office: (970) 494-7478.

15 See footnote 1 in § 3.111(e)(8).

records of the animal; and a discussion of each animal with an animal care staff member familiar with the animal.

(2) The attending veterinarian shall observe an interactive swim session at the SWTD site at least once each month.

(3) The attending veterinarian shall conduct a complete physical examination of each cetacean at least once every 6 months. The examination shall include a profile of the cetacean, including the cetacean's identification (name and/or number, sex, and age), weight,[16] length, axillary girth, appetite, and behavior. The attending veterinarian shall also conduct a general examination to evaluate body condition, skin, eyes, mouth, blow hole and cardio-respiratory system, genitalia, and feces (gastrointestinal status). The examination shall also include a complete blood count and serum chemistry analysis. Fecal and blow hole smears shall be obtained for cytology and parasite evaluation.

(4) The attending veterinarian, during the monthly site visit, shall record the nutritional and reproductive status of each cetacean (i.e., whether in an active breeding program, pregnant, or nursing).

(5) The attending veterinarian shall examine water quality records and provide a written assessment, to remain at the SWTD site for at least 3 years, of the overall water quality during the preceding month. Such records shall be made available to an APHIS official upon request during inspection.

(6) In the event that a cetacean dies, complete necropsy results, including all appropriate histopathology, shall be recorded in the cetacean's individual file and shall be made available to APHIS officials during facility inspections, or as requested by APHIS. The necropsy shall be performed within 48 hours of the cetacean's death, by a veterinarian experienced in marine mammal necropsies. If the necropsy is not to be performed within 3 hours of the discovery of the cetacean's death, the cetacean shall be refrigerated until necropsy. Written results of the necropsy shall be available in the cetacean's individual file within 7 days after death for gross pathology and within 45 days after death for histopathology.

(Approved by the Office of Management and Budget under control numbers 0579-0036 and 0579-0115)

[63 FR 47148, Sept. 4, 1998]
Effective Date Note: At 64 FR 15920, Apr. 2, 1999, § 3.111 was suspended, effective Apr. 2, 1999.

---

16 Weight may be measured either by scale or calculated using the following formulae:
   Females: Natural log of body mass = -8.44 + 1.34(natural log of girth) + 1.28(natural log of standard length).
   Males: Natural log of body mass = -10.3 + 1.62(natural log of girth) + 1.38(natural log of standard length).

TRANSPORTATION STANDARDS

§ 3.112 - Consignments to carriers and intermediate handlers.

(a) Carriers and intermediate handlers shall not accept any marine mammal that is presented by any dealer, research facility, exhibitor, operator of an auction sale, or other person, or any department, agency, or instrumentality of the United States or any State or local government for shipment, in commerce, more than 4 hours prior to the scheduled departure of the primary conveyance on which it is to be transported, and that is not accompanied by a health certificate signed by the attending veterinarian stating that the animal was examined within the prior 10 days and found to be in acceptable health for transport: *Provided, however,* That the carrier or intermediate handler and any dealer, research facility, exhibitor, operator of an auction sale, or other person, or any department, agency, or instrumentality of the United States or any State or local government may mutually agree to extend the time of acceptance to not more than 6 hours if specific prior scheduling of the animal shipment to destination has been made.

(b) Any carrier or intermediate handler shall only accept for transportation or transport, in commerce, any marine mammal in a primary transport enclosure that conforms to the requirements in § 3.113 of this subpart: *Provided, however,* That any carrier or intermediate handler may accept for transportation or transport, in commerce, any marine mammal consigned by any department, agency, or instrumentality of the United States having laboratory animal facilities or exhibiting animals or any licensed or registered dealer, research facility, exhibitor, or operator of an auction sale if the consignor furnishes to the carrier or intermediate handler a certificate, signed by the consignor, stating that the primary transport enclosure complies with § 3.113 of this subpart, unless such primary transport enclosure is obviously defective or damaged and it is apparent that it cannot reasonably be expected to contain the marine mammal without causing suffering or injury to the marine mammal. A copy of any such certificate must accompany the shipment to destination. The certificate must include at least the following information:

(1) Name and address of the consignor;

(2) The number, age, and sex of animals in the primary transport enclosure(s);

(3) A certifying statement (e.g., "I hereby certify that the—(number) primary transport enclosure(s) that are used to transport the animal(s) in this shipment complies (comply) with USDA standards for primary transport enclosures (9 CFR part 3)."); and

(4) The signature of the consignor, and date.

PART 3
Subpart E

(c) Carriers or intermediate handlers whose facilities fail to maintain a temperature within the range of 7.2 °C (45 °F) to 23.9 °C (75 °F) allowed by § 3.117 of this subpart may accept for transportation or transport, in commerce, any marine mammal consigned by any department, agency, or instrumentality of the United States or of any State or local government, or by any person (including any licensee or registrant under the Act, as well as any private individual) if the consignor furnishes to the carrier or intermediate handler a certificate executed by the attending veterinarian on a specified date that is not more than 10 days prior to delivery of the animal for transportation in commerce, stating that the marine mammal is acclimated to a specific air temperature range lower or higher than those prescribed in §§ 3.117 and 3.118. A copy of the certificate must accompany the shipment to destination. The certificate must include at least the following information:

(1) Name and address of the consignor;

(2) The number, age, and sex of animals in the shipment;

(3) A certifying statement (e.g., "I hereby certify that the animal(s) in this shipment is (are), to the best of my knowledge, acclimated to an air temperature range of ____"); and

(4) The signature of the attending veterinarian and the date.

(d) Carriers and intermediate handlers must attempt to notify the consignee (receiving party) at least once in every 6-hour period following the arrival of any marine mammals at the animal holding area of the terminal cargo facility. The time, date, and method of each attempted notification and the final notification to the consignee and the name of the person notifying the consignee must be recorded on the copy of the shipping document retained by the carrier or intermediate handler and on a copy of the shipping document accompanying the animal shipment.

[66 FR 254, Jan. 3, 2001]

§ 3.113 - Primary enclosures used to transport marine mammals.

No dealer, research facility, exhibitor, or operator of an auction sale shall offer for transportation or transport, in commerce, any marine mammal in a primary enclosure that does not conform to the following requirements:

(a) Primary enclosures that are used to transport marine mammals other than cetaceans and sirenians must:

(1) Be constructed from materials of sufficient structural strength to contain the marine mammals;

(2) Be constructed from material that is durable, nontoxic, and cannot be chewed and/or swallowed;

(3) Be able to withstand the normal rigors of transportation;

(4) Have interiors that are free from any protrusions or hazardous openings that could be injurious to the marine mammals contained within;

(5) Be constructed so that no parts of the contained marine mammals are exposed to the outside of the enclosures in any way that may cause injury to the animals or to persons who are nearby or who handle the enclosures;

(6) Have openings that provide access into the enclosures and are secured with locking devices of a type that cannot be accidentally opened;

(7) Have such openings located in a manner that makes them easily accessible at all times for emergency removal and potential treatment of any live marine mammal contained within;

(8) Have air inlets at heights that will provide cross ventilation at all levels (particularly when the marine mammals are in a prone position), are located on all four sides of the enclosures, and cover not less than 20 percent of the total surface area of each side of the enclosures;

(9) Have projecting rims or other devices placed on any ends and sides of the enclosures that have ventilation openings so that there is a minimum air circulation space of 7.6 centimeters (3.0 inches) between the enclosures and any adjacent cargo or conveyance wall;

(10) Be constructed so as to provide sufficient air circulation space to maintain the temperature limits set forth in this subpart; and

(11) Be equipped with adequate handholds or other devices on the exterior of the enclosures to enable them to be lifted without unnecessary tilting and to ensure that the persons handling the enclosures will not come in contact with any marine mammal contained inside.

(b) Straps, slings, harnesses, or other devices used for body support or restraint, when transporting marine mammals such as cetaceans and sirenians must:

(1) Be designed so as not to prevent access to the marine mammals by attendants for the purpose of administering in-transit care;

(2) Be equipped with special padding to prevent trauma or injury at critical weight pressure points on the body of the marine mammals; and

(3) Be capable of keeping the animals from thrashing about and causing injury to themselves or their attendants, and yet be adequately designed so as not to cause injury to the animals.

(c) Primary enclosures used to transport marine mammals must be large enough to assure that:

(1) In the case of pinnipeds, polar bears, and sea otters, each animal has sufficient space to turn about freely in a stance whereby all four feet or flippers are on the floor and the animal can sit in an upright position and lie in a natural position;

(2) In the case of cetaceans and sirenians, each animal has sufficient space for support of its body in slings, harnesses, or other supporting devices, if used (as prescribed in paragraph (b) of this section), without

PART 3
Subpart E

223

causing injury to such cetaceans or sirenians due to contact with the primary transport enclosure: *Provided, however,* That animals may be restricted in their movements according to professionally accepted standards when such freedom of movement would constitute a danger to the animals, their handlers, or other persons.

(d) Marine mammals transported in the same primary enclosure must be of the same species and maintained in compatible groups. Marine mammals that have not reached puberty may not be transported in the same primary enclosure with adult marine mammals other than their dams. Socially dependent animals (e.g., sibling, dam, and other members of a family group) must be allowed visual and olfactory contact whenever reasonable. Female marine mammals may not be transported in the same primary enclosure with any mature male marine mammals.

(e) Primary enclosures used to transport marine mammals as provided in this section must have solid bottoms to prevent leakage in shipment and must be cleaned and sanitized in a manner prescribed in § 3.107 of this subpart, if previously used. Within the primary enclosures used to transport marine mammals, the animals will be maintained on sturdy, rigid, solid floors with adequate drainage.

(f) Primary enclosures used to transport marine mammals, except where such primary enclosures are permanently affixed in the animal cargo space of the primary conveyance, must be clearly marked on top (when present) and on at least one side, or on all sides whenever possible, with the words "Live Animal" or "Wild Animal" in letters not less than 2.5 centimeters (1 inch) in height, and with arrows or other markings to indicate the correct upright position of the container.

(g) Documents accompanying the shipment must be attached in an easily accessible manner to the outside of a primary enclosure that is part of such shipment or be in the possession of the shipping attendant.

(h) When a primary transport enclosure is permanently affixed within the animal cargo space of the primary conveyance so that the front opening is the only source of ventilation for such primary enclosure, the front opening must open directly to the outside or to an unobstructed aisle or passageway within the primary conveyance. Such front ventilation opening must be at least 90 percent of the total surface area of the front wall of the primary enclosure and covered with bars, wire mesh, or smooth expanded metal.

PART 3
Subpart E

[66 FR 255, Jan. 3, 2001]

§ 3.114 - Primary conveyances (motor vehicle, rail, air and marine).

(a) The animal cargo space of primary conveyances used in transporting live marine mammals must be constructed in a manner that will protect the health and assure the safety and comfort of the marine mammals contained within at all times. All primary conveyances used must be sufficiently temperature-controlled to provide an appropriate environmental temperature for the species involved and to provide for the safety and comfort of the marine mammal, or other appropriate safeguards (such as, but not limited to, cooling the animal with cold water, adding ice to water-filled enclosures, and use of fans) must be employed to maintain the animal at an appropriate temperature.

(b) The animal cargo space must be constructed and maintained in a manner that will prevent the ingress of engine exhaust fumes and gases in excess of that ordinarily contained in the passenger compartments.

(c) Marine mammals must only be placed in animal cargo spaces that have a supply of air sufficient for each live animal contained within. Primary transport enclosures must be positioned in the animal cargo spaces of primary conveyances in such a manner that each marine mammal contained within will have access to sufficient air.

(d) Primary transport enclosures must be positioned in primary conveyances in such a manner that, in an emergency, the live marine mammals can be removed from the conveyances as soon as possible.

(e) The interiors of animal cargo spaces in primary conveyances must be kept clean.

(f) Live marine mammals must not knowingly be transported with any material, substance, or device that may be injurious to the health and well-being of the marine mammals unless proper precaution is taken to prevent such injury.

(g) Adequate lighting must be available for marine mammal attendants to properly inspect the animals at any time. If such lighting is not provided by the carrier, provisions must be made by the shipper to supply such lighting.

[66 FR 255, Jan. 3, 2001]

§ 3.115 - Food and drinking water requirements.
(a) Those marine mammals that require drinking water must be offered potable water within 4 hours of being placed in the primary transport enclosure for transport in commerce. Marine mammals must be provided water as often as necessary and appropriate for the species involved to prevent dehydration, which would jeopardize the good health and well-being of the animals.

(b) Marine mammals being transported in commerce must be offered food as often as necessary and appropriate for the species involved or as determined by the attending veterinarian.

[66 FR 256, Jan. 3, 2001]

PART 3
Subpart E

§ 3.116 - Care in transit.

(a) A licensed veterinarian, employee, and/or attendant of the shipper or receiver of any marine mammal being transported, in commerce, knowledgeable and experienced in the area of marine mammal care and transport, must accompany all marine mammals during periods of transportation to provide for their good health and well-being, to observe such marine mammals to determine whether they need veterinary care, and to obtain any needed veterinary care as soon as possible. Any transport of greater than 2 hours duration requires a transport plan approved by the attending veterinarian that will include the specification of the necessity of the presence of a veterinarian during the transport. If the attending veterinarian does not accompany the animal, communication with the veterinarian must be maintained in accordance with § § 2.33(b)(3) and 2.40(b)(3) of this chapter.

(b) The following marine mammals may be transported in commerce only when the transport of such marine mammals has been determined to be appropriate by the attending veterinarian:

(1) A pregnant animal in the last half of pregnancy;

(2) A dependent unweaned young animal;

(3) A nursing mother with young; or

(4) An animal with a medical condition requiring veterinary care, that would be compromised by transport. The attending veterinarian must note on the accompanying health certificate the existence of any of the above conditions. The attending veterinarian must also determine whether a veterinarian should accompany such marine mammals during transport.

(c) Carriers must inform the crew as to the presence of the marine mammals on board the craft, inform the individual accompanying the marine mammals of any unexpected delays as soon as they become known, and accommodate, except as precluded by safety considerations, requests by the shipper or his agent to provide access to the animals or take other necessary actions for the welfare of the animals if a delay occurs.

(d) A sufficient number of employees or attendants of the shipper or receiver of cetaceans or sirenians being transported, in commerce, must provide for such cetaceans and sirenians during periods of transport by:

(1) Keeping the skin moist or preventing the drying of the skin by such methods as intermittent spraying of water or application of a nontoxic emollient;

(2) Assuring that the pectoral flippers are allowed freedom of movement at all times;

(3) Making adjustments in the position of the marine mammals when necessary to prevent necrosis of the skin at weight pressure points;

PART 3
Subpart E

(4) Keeping the animal cooled and/or warmed sufficiently to prevent overheating, hypothermia, or temperature related stress; and

(5) Calming the marine mammals to avoid struggling, thrashing, and other unnecessary activity that may cause overheating or physical trauma.

(e) A sufficient number of employees or attendants of the shipper or receiver of pinnipeds or polar bears being transported, in commerce, must provide for such pinnipeds and polar bears during periods of transport by:

(1) Keeping the animal cooled and/or warmed sufficiently to prevent overheating, hypothermia, or temperature related stress; and

(2) Calming the marine mammals to avoid struggling, thrashing, and other unnecessary activity that may cause overheating or physical trauma.

(f) Sea otters must be transported in primary enclosures that contain false floors through which water and waste freely pass to keep the interior of the transport unit free from waste materials. Moisture must be provided by water sprayers or ice during transport.

(g) Marine mammals may be removed from their primary transport enclosures only by the attendants or other persons capable of handling such mammals safely.

[66 FR 256, Jan. 3, 2001]

§ 3.117 - Terminal facilities.

Carriers and intermediate handlers must not commingle marine mammal shipments with inanimate cargo. All animal holding areas of a terminal facility of any carrier or intermediate handler where marine mammal shipments are maintained must be cleaned and sanitized in a manner prescribed in § 3.107 of this subpart to minimize health and disease hazards. An effective program for the control of insects, ectoparasites, and avian and mammalian pests must be established and maintained for all animal holding areas. Any animal holding area containing marine mammals must be ventilated with fresh air or air circulated by means of fans, blowers, or an air conditioning system so as to minimize drafts, odors, and moisture condensation. Auxiliary ventilation, such as exhaust fans and vents or fans or blowers or air conditioning must be used for any animal holding area containing marine mammals when the air temperature within such animal holding area is 23.9 °C (75 °F) or higher. The air temperature around any marine mammal in any animal holding area must not be allowed to fall below 7.2 °C (45 °F). The air temperature around any polar bear must not be allowed to exceed 29.5 °C (85 °F) at any time and no polar bear may be subjected to surrounding air temperatures that exceed 23.9 °C (75 °F) for more than 4 hours at any time. The ambient temperature must be measured in the animal holding area upon arrival of the shipment by the attendant,

PART 3
Subpart E

carrier, or intermediate handler. The ambient temperature must be measured halfway up the outside of the primary transport enclosure at a distance from the external wall of the primary transport enclosure not to exceed 0.91 meters (3 feet).

[66 FR 256, Jan. 3, 2001]

§ 3.118 - Handling.

(a) Carriers and intermediate handlers moving marine mammals from the animal holding area of the terminal facility to the primary conveyance or from the primary conveyance to the animal holding area of the terminal facility must provide the following:

(1) *Movement of animals as expeditiously as possible.*

(2) *Shelter from overheating and direct sunlight.* When sunlight is likely to cause overheating, sunburn, or discomfort, sufficient shade must be provided to protect the marine mammals. Marine mammals must not be subjected to surrounding air temperatures that exceed 23.9 °C (75 °F) unless accompanied by an acclimation certificate in accordance with § 3.112 of this subpart. The temperature must be measured and read within or immediately adjacent to the primary transport enclosure.

(3) *Shelter from cold weather.* Marine mammals must be provided with species appropriate protection against cold weather, and such marine mammals must not be subjected to surrounding air temperatures that fall below 7.2 °C (45 °F) unless accompanied by an acclimation certificate in accordance with § 3.112 of this subpart. The temperature must be measured and read within or immediately adjacent to the primary transport enclosure.

(b) Care must be exercised to avoid handling of the primary transport enclosure in a manner that may cause physical harm or distress to the marine mammal contained within.

(c) Enclosures used to transport any marine mammal must not be tossed, dropped, or needlessly tilted and must not be stacked unless properly secured.

PART 3
Subpart E

[66 FR 257, Jan. 3, 2001]

Subpart F – Specifications for the Humane Handling, Care, Treatment, and Transportation of Warmblooded Animals Other Than Dogs, Cats, Rabbits, Hamsters, Guinea Pigs, Nonhuman Primates, and Marine Mammals

Source: 36 FR 24925, Dec. 24, 1971, unless otherwise noted. Redesignated at 44 FR 36874, July 22, 1979.

FACILITIES AND OPERATING STANDARDS

§ 3.125 - Facilities, general.

   (a) *Structural strength.* The facility must be constructed of such material and of such strength as appropriate for the animals involved. The indoor and outdoor housing facilities shall be structurally sound and shall be maintained in good repair to protect the animals from injury and to contain the animals.

   (b) *Water and power.* Reliable and adequate electric power, if required to comply with other provisions of this subpart, and adequate potable water shall be available on the premises.

   (c) *Storage.* Supplies of food and bedding shall be stored in facilities which adequately protect such supplies against deterioration, molding, or contamination by vermin. Refrigeration shall be provided for supplies of perishable food.

   (d) *Waste disposal.* Provision shall be made for the removal and disposal of animal and food wastes, bedding, dead animals, trash and debris. Disposal facilities shall be so provided and operated as to minimize vermin infestation, odors, and disease hazards. The disposal facilities and any disposal of animal and food wastes, bedding, dead animals, trash, and debris shall comply with applicable Federal, State, and local laws and regulations relating to pollution control or the protection of the environment.

   (e) *Washroom and sinks.* Facilities, such as washrooms, basins, showers, or sinks, shall be provided to maintain cleanliness among animal caretakers.

[36 FR 24925, Dec. 24, 1971. Redesignated at 44 FR 36874, June 22, 1979, and amended at 44 FR 63492, Nov. 2, 1979]

§ 3.126 - Facilities, indoor.

   (a) *Ambient temperatures.* Temperature in indoor housing facilities shall be sufficiently regulated by heating or cooling to protect the animals from the extremes of temperature, to provide for their health and to prevent their discomfort. The ambient temperature shall not be allowed to fall below nor rise above temperatures compatible with the health and comfort of the animal.

PART 3

Subpart F

(b) *Ventilation.* Indoor housing facilities shall be adequately ventilated by natural or mechanical means to provide for the health and to prevent discomfort of the animals at all times. Such facilities shall be provided with fresh air either by means of windows, doors, vents, fans, or air-conditioning and shall be ventilated so as to minimize drafts, odors, and moisture condensation.

(c) *Lighting.* Indoor housing facilities shall have ample lighting, by natural or artificial means, or both, of good quality, distribution, and duration as appropriate for the species involved. Such lighting shall be uniformly distributed and of sufficient intensity to permit routine inspection and cleaning. Lighting of primary enclosures shall be designed to protect the animals from excessive illumination.

(d) *Drainage.* A suitable sanitary method shall be provided to eliminate rapidly, excess water from indoor housing facilities. If drains are used, they shall be properly constructed and kept in good repair to avoid foul odors and installed so as to prevent any backup of sewage. The method of drainage shall comply with applicable Federal, State, and local laws and regulations relating to pollution control or the protection of the environment.

§ 3.127 - Facilities, outdoor.

(a) *Shelter from sunlight.* When sunlight is likely to cause overheating or discomfort of the animals, sufficient shade by natural or artificial means shall be provided to allow all animals kept outdoors to protect themselves from direct sunlight.

(b) *Shelter from inclement weather.* Natural or artificial shelter appropriate to the local climatic conditions for the species concerned shall be provided for all animals kept outdoors to afford them protection and to prevent discomfort to such animals. Individual animals shall be acclimated before they are exposed to the extremes of the individual climate.

(c) *Drainage.* A suitable method shall be provided to rapidly eliminate excess water. The method of drainage shall comply with applicable Federal, State, and local laws and regulations relating to pollution control or the protection of the environment.

(d) *Perimeter fence.* On or after May 17, 2000, all outdoor housing facilities (i.e., facilities not entirely indoors) must be enclosed by a perimeter fence that is of sufficient height to keep animals and unauthorized persons out. Fences less than 8 feet high for potentially dangerous animals, such as, but not limited to, large felines (e.g., lions, tigers, leopards, cougars, etc.), bears, wolves, rhinoceros, and elephants, or less than 6 feet high for other animals must be approved in writing by the Administrator. The fence must be constructed so that it protects the animals in the facility by restricting animals and unauthorized persons from going through it or under it and

PART 3
Subpart F

having contact with the animals in the facility, and so that it can function as a secondary containment system for the animals in the facility. It must be of sufficient distance from the outside of the primary enclosure to prevent physical contact between animals inside the enclosure and animals or persons outside the perimeter fence. Such fences less than 3 feet in distance from the primary enclosure must be approved in writing by the Administrator. A perimeter fence is not required:

(1) Where the outside walls of the primary enclosure are made of sturdy, durable material, which may include certain types of concrete, wood, plastic, metal, or glass, and are high enough and constructed in a manner that restricts entry by animals and unauthorized persons and the Administrator gives written approval; or

(2) Where the outdoor housing facility is protected by an effective natural barrier that restricts the animals to the facility and restricts entry by animals and unauthorized persons and the Administrator gives written approval; or

(3) Where appropriate alternative security measures are employed and the Administrator gives written approval; or

(4) For traveling facilities where appropriate alternative security measures are employed; or

(5) Where the outdoor housing facility houses only farm animals, such as, but not limited to, cows, sheep, goats, pigs, horses (for regulated purposes), or donkeys, and the facility has in place effective and customary containment and security measures.

[36 FR 24925, Dec. 24, 1971. Redesignated at 44 FR 36874, July 22, 1979, as amended at 64 FR 56147, Oct. 18, 1999; 65 FR 70770, Nov. 28, 2000]

§ 3.128 - Space requirements.

Enclosures shall be constructed and maintained so as to provide sufficient space to allow each animal to make normal postural and social adjustments with adequate freedom of movement. Inadequate space may be indicated by evidence of malnutrition, poor condition, debility, stress, or abnormal behavior patterns.

ANIMAL HEALTH AND HUSBANDRY STANDARDS

§ 3.129 - Feeding.

(a) The food shall be wholesome, palatable, and free from contamination and of sufficient quantity and nutritive value to maintain all animals in good health. The diet shall be prepared with consideration for the age, species, condition, size, and type of the animal. Animals shall be fed at least once a

PART 3
Subpart F

day except as dictated by hibernation, veterinary treatment, normal fasts, or other professionally accepted practices.

(b) Food, and food receptacles, if used, shall be sufficient in quantity and located so as to be accessible to all animals in the enclosure and shall be placed so as to minimize contamination. Food receptacles shall be kept clean and sanitary at all times. If self-feeders are used, adequate measures shall be taken to prevent molding, contamination, and deterioration or caking of food.

§ 3.130 - Watering.

If potable water is not accessible to the animals at all times, it must be provided as often as necessary for the health and comfort of the animal. Frequency of watering shall consider age, species, condition, size, and type of the animal. All water receptacles shall be kept clean and sanitary.

§ 3.131 - Sanitation.

(a) *Cleaning of enclosures.* Excreta shall be removed from primary enclosures as often as necessary to prevent contamination of the animals contained therein and to minimize disease hazards and to reduce odors. When enclosures are cleaned by hosing or flushing, adequate measures shall be taken to protect the animals confined in such enclosures from being directly sprayed with the stream of water or wetted involuntarily.

(b) *Sanitation of enclosures.* Subsequent to the presence of an animal with an infectious or transmissible disease, cages, rooms, and hard-surfaced pens or runs shall be sanitized either by washing them with hot water (180 F. at source) and soap or detergent, as in a mechanical washer, or by washing all soiled surfaces with a detergent solution followed by a safe and effective disinfectant, or by cleaning all soiled surfaces with saturated live steam under pressure. Pens or runs using gravel, sand, or dirt, shall be sanitized when necessary as directed by the attending veterinarian.

(c) *Housekeeping.* Premises (buildings and grounds) shall be kept clean and in good repair in order to protect the animals from injury and to facilitate the prescribed husbandry practices set forth in this subpart. Accumulations of trash shall be placed in designated areas and cleared as necessary to protect the health of the animals.

(d) *Pest control.* A safe and effective program for the control of insects, ectoparasites, and avian and mammalian pests shall be established and maintained.

PART 3
Subpart F

§ 3.132 - Employees.

A sufficient number of adequately trained employees shall be utilized to maintain the professionally acceptable level of husbandry practices set forth in this subpart. Such practices shall be under a supervisor who has a background in animal care.

§ 3.133 - Separation.

Animals housed in the same primary enclosure must be compatible. Animals shall not be housed near animals that interfere with their health or cause them discomfort.

§§ 3.134-3.135 - [Reserved]

TRANSPORTATION STANDARDS

Source: Sections 3.136 through 3.142 appear at 42 FR 31569, June 21, 1977, unless otherwise noted. Redesignated at 44 FR 36874, July 22, 1979.

§ 3.136 - Consignments to carriers and intermediate handlers.

(a) Carriers and intermediate handlers shall not accept any live animals presented by any dealer, research facility, exhibitor, operator of an auction sale, or other person, or any department, agency, or instrumentality of the United States or any State or local government for shipment, in commerce, more than 4 hours prior to the scheduled departure of the primary conveyance on which it is to be transported: *Provided, however,* That the carrier or intermediate handler and any dealer, research facility, exhibitor, operator of an auction sale, or other person, or any department, agency, or instrumentality of the United States or any State or local government may mutually agree to extend the time of acceptance to not more than 6 hours if specific prior scheduling of the animal shipment to destination has been made.

(b) Any carrier or intermediate handler shall only accept for transportation or transport, in commerce, any live animal in a primary enclosure which conforms to the requirements set forth in § 3.137 of the standards: *Provided, however,* That any carrier or intermediate handler may accept for transportation or transport, in commerce, any live animal consigned by any department, agency, or instrumentality of the United States having laboratory animal facilities or exhibiting animals or any licensed or registered dealer, research facility, exhibitor, or operator of an auction sale if the consignor furnishes to the carrier or intermediate handler a certificate, signed by the consignor, stating that the primary enclosure complies with § 3.137 of the standards, unless such primary enclosure is obviously defective or damaged and it is apparent that it cannot reasonably be expected to contain the live animal without causing suffering or injury to such live animal. A copy of such

PART 3
Subpart F

certificate shall accompany the shipment to destination. The certificate shall include at least the following information:

(1) Name and address of the consignor;

(2) The number of animals in the primary enclosure(s);

(3) A certifying statement (e.g., "I hereby certify that the __ (number) primary enclosure(s) which are used to transport the animal(s) in this shipment complies (comply) with USDA standards for primary enclosures (9 CFR part 3)."); and

(4) The signature of the consignor, and date.

(c) Carriers or intermediate handlers whose facilities fail to meet the minimum temperature allowed by the standards may accept for transportation or transport, in commerce, any live animal consigned by any department, agency, or instrumentality of the United States or of any State or local government, or by any person (including any licensee or registrant under the Act, as well as any private individual) if the consignor furnishes to the carrier or intermediate handler a certificate executed by a veterinarian accredited by this Department pursuant to part 160 of this title on a specified date which shall not be more than 10 days prior to delivery of such animal for transportation in commerce, stating that such live animal is acclimated to air temperatures lower than those prescribed in §§ 3.141 and 3.142. A copy of such certificate shall accompany the shipment to destination. The certificate shall include at least the following information:

(1) Name and address of the consignor;

(2) The number of animals in the shipment;

(3) A certifying statement (e.g., "I hereby certify that the animal(s) in this shipment is (are), to the best of my knowledge, acclimated to air temperatures lower than 7.2 °C. (45 °F.)"); and

(4) The signature of the USDA accredited veterinarian, assigned accreditation number, and date.

(d) Carriers and intermediate handlers shall attempt to notify the consignee at least once in every 6 hour period following the arrival of any live animals at the animal holding area of the terminal cargo facility. The time, date, and method of each attempted notification and the final notification to the consignee and the name of the person notifying the consignee shall be recorded on the copy of the shipping document retained by the carrier or intermediate handler and on a copy of the shipping document accompanying the animal shipment.

PART 3
Subpart F

[42 FR 31569, June 21, 1977, as amended at 43 FR 21166, May 16, 1978. Redesignated at 44 FR 36874, July 22, 1979, and amended at 44 FR 63493, Nov. 2, 1979]

§ 3.137 - Primary enclosures used to transport live animals.

No dealer, research facility, exhibitor, or operator of an auction sale shall offer for transportation or transport, in commerce, any live animal in a primary enclosure which does not conform to the following requirements:

(a) Primary enclosures, such as compartments, transport cages, cartons, or crates, used to transport live animals shall be constructed in such a manner that **(1)** the structural strength of the enclosure shall be sufficient to contain the live animals and to withstand the normal rigors of transportation; **(2)** the interior of the enclosure shall be free from any protrusions that could be injurious to the live animals contained therein; **(3)** the openings of such enclosures are easily accessible at all times for emergency removal of the live animals; **(4)** except as provided in paragraph (g) of this section, there are ventilation openings located on two opposite walls of the primary enclosure and the ventilation openings on each such wall shall be at least 16 percent of the total surface area of each such wall, or there are ventilation openings located on all four walls of the primary enclosure and the ventilation openings on each such wall shall be at least 8 percent of the total surface area of each such wall: *Provided, however,* That at least one-third of the total minimum area required for ventilation of the primary enclosure shall be located on the lower one-half of the primary enclosure and at least one-third of the total minimum area required for ventilation of the primary enclosure shall be located on the upper one-half of the primary enclosure; **(5)** except as provided in paragraph (g) of this section, projecting rims or other devices shall be on the exterior of the outside walls with any ventilation openings to prevent obstruction of the ventilation openings and to provide a minimum air circulation space of 1.9 centimeters (.75 inch) between the primary enclosure and any adjacent cargo or conveyance wall; and **(6)** except as provided in paragraph (g) of this section, adequate handholds or other devices for lifting shall be provided on the exterior of the primary enclosure to enable the primary enclosure to be lifted without tilting and to ensure that the person handling the primary enclosure will not be in contact with the animal.

(b) Live animals transported in the same primary enclosure shall be of the same species and maintained in compatible groups. Live animals that have not reached puberty shall not be transported in the same primary enclosure with adult animals other than their dams. Socially dependent animals (e.g., sibling, dam, and other members of a family group) must be allowed visual and olfactory contact. Any female animal in season (estrus) shall not be transported in the same primary enclosure with any male animal.

(c) Primary enclosures used to transport live animals shall be large enough to ensure that each animal contained therein has sufficient space to turn about freely and to make normal postural adjustments: *Provided, however,* That certain species may be restricted in their movements according to

PART 3
Subpart F

professionally acceptable standards when such freedom of movement would constitute a danger to the animals, their handlers, or other persons.

(d) Primary enclosures used to transport live animals as provided in this section shall have solid bottoms to prevent leakage in shipment and still be cleaned and sanitized in a manner prescribed in § 3.131 of the standards, if previously used. Such primary enclosures shall contain clean litter of a suitable absorbent material, which is safe and nontoxic to the live animals contained therein, in sufficient quantity to absorb and cover excreta, unless the animals are on wire or other nonsolid floors.

(e) Primary enclosures used to transport live animals, except where such primary enclosures are permanently affixed in the animal cargo space of the primary conveyance, shall be clearly marked on top and on one or more sides with the words "Live Animal" or "Wild Animal", whichever is appropriate, in letters not less than 2.5 centimeters (1 inch) in height, and with arrows or other markings to indicate the correct upright position of the container.

(f) Documents accompanying the shipment shall be attached in an easily accessible manner to the outside of a primary enclosure which is part of such shipment.

(g) When a primary enclosure is permanently affixed within the animal cargo space of the primary conveyance so that the front opening is the only source of ventilation for such primary enclosure, the front opening shall open directly to the outside or to an unobstructed aisle or passageway within the primary conveyance. Such front ventilation opening shall be at least 90 percent of the total surface area of the front wall of the primary enclosure and covered with bars, wire mesh or smooth expanded metal.

[42 FR 31569, June 21, 1977, as amended at 43 FR 21166, May 16, 1978. Redesignated at 44 FR 36874, July 22, 1979]

§ 3.138 - Primary conveyances (motor vehicle, rail, air, and marine).

(a) The animal cargo space of primary conveyances used in transporting live animals shall be designed and constructed to protect the health, and ensure the safety and comfort of the live animals contained therein at all times.

(b) The animal cargo space shall be constructed and maintained in a manner to prevent the ingress of engine exhaust fumes and gases from the primary conveyance during transportation in commerce.

PART 3
Subpart F

(c) No live animal shall be placed in an animal cargo space that does not have a supply of air sufficient for normal breathing for each live animal contained therein, and the primary enclosures shall be positioned in the animal cargo space in such a manner that each live animal has access to sufficient air for normal breathing.

(d) Primary enclosures shall be positioned in the primary conveyance in such a manner that in an emergency the live animals can be removed from the primary conveyance as soon as possible.

(e) The interior of the animal cargo space shall be kept clean.

(f) Live animals shall not be transported with any material, substance (e.g., dry ice) or device which may reasonably be expected to be injurious to the health and well-being of the animals unless proper precaution is taken to prevent such injury.

§ 3.139 - Food and water requirements.

(a) All live animals shall be offered potable water within 4 hours prior to being transported in commerce. Dealers, exhibitors, research facilities and operators of auction sales shall provide potable water to all live animals transported in their own primary conveyance at least every 12 hours after such transportation is initiated, and carriers and intermediate handlers shall provide potable water to all live animals at least every 12 hours after acceptance for transportation in commerce: *Provided, however,* That except as directed by hibernation, veterinary treatment or other professionally accepted practices, those live animals which, by common accepted practices, require watering more frequently shall be so watered.

(b) Each live animal shall be fed at least once in each 24 hour period, except as directed by hibernation, veterinary treatment, normal fasts, or other professionally accepted practices. Those live animals which, by common accepted practice, require feeding more frequently shall be so fed.

(c) A sufficient quantity of food and water shall accompany the live animal to provide food and water for such animals for a period of at least 24 hours, except as directed by hibernation, veterinary treatment, normal fasts, and other professionally accepted practices.

(d) Any dealer, research facility, exhibitor or operator of an auction sale offering any live animal to any carrier or intermediate handler for transportation in commerce shall affix to the outside of the primary enclosure used for transporting such live animal, written instructions concerning the food and water requirements of such animal while being so transported.

(e) No carrier or intermediate handler shall accept any live animals for transportation in commerce unless written instructions concerning the food and water requirements of such animal while being so transported is affixed to the outside of its primary enclosure.

§ 3.140 - Care in transit.

(a) During surface transportation, it shall be the responsibility of the driver or other employee to visually observe the live animals as frequently as circumstances may dictate, but not less than once every 4 hours, to assure

PART 3
Subpart F

237

that they are receiving sufficient air for normal breathing, their ambient temperatures are within the prescribed limits, all other applicable standards are being complied with and to determine whether any of the live animals are in obvious physical distress and to provide any needed veterinary care as soon as possible. When transported by air, live animals shall be visually observed by the carrier as frequently as circumstances may dictate, but not less than once every 4 hours, if the animal cargo space is accessible during flight. If the animal cargo space is not accessible during flight, the carrier shall visually observe the live animals whenever loaded and unloaded and whenever the animal cargo space is otherwise accessible to assure that they are receiving sufficient air for normal breathing, their ambient temperatures are within the prescribed limits, all other applicable standards are being complied with and to determine whether any such live animals are in obvious physical distress. The carrier shall provide any needed veterinary care as soon as possible. No animal in obvious physical distress shall be transported in commerce.

(b) Wild or otherwise dangerous animals shall not be taken from their primary enclosure except under extreme emergency conditions: *Provided, however,* That a temporary primary enclosure may be used, if available, and such temporary primary enclosure is structurally strong enough to prevent the escape of the animal. During the course of transportation, in commerce, live animals shall not be removed from their primary enclosures unless placed in other primary enclosures or facilities conforming to the requirements provided in this subpart.

§ 3.141 - Terminal facilities.

Carriers and intermediate handlers shall not commingle live animal shipments with inanimate cargo. All animal holding areas of a terminal facility of any carrier or intermediate handler wherein live animal shipments are maintained shall be cleaned and sanitized in a manner prescribed in § 3.141 of the standards often enough to prevent an accumulation of debris or excreta, to minimize vermin infestation and to prevent a disease hazard. An effective program for the control of insects, ectoparasites, and avian and mammalian pests shall be established and maintained for all animal holding areas. Any animal holding area containing live animals shall be provided with fresh air by means of windows, doors vents, or air conditioning and may be ventilated or air circulated by means of fans, blowers, or an air conditioning system so as to minimize drafts, odors, and moisture condensation. Auxiliary ventilation, such as exhaust fans and vents or fans or blowers or air conditioning shall be used for any animal holding area containing live animals when the air temperature within such animal holding area is 23.9 °C. (75.°F.) or higher. The air temperature around any live animal in any

PART 3
Subpart F

animal holding area shall not be allowed to fall below 7.2 °C. (45 °F.) nor be allowed to exceed 29.5 °C. (85 °F.) at any time: *Provided, however,* That no live animal shall be subjected to surrounding air temperatures which exceed 23.9 °C. (75 °F.) for more than 4 hours at any time. To ascertain compliance with the provisions of this paragraph, the air temperature around any live animal shall be measured and read outside the primary enclosure which contains such animal at a distance not to exceed .91 meters (3 feet) from any one of the external walls of the primary enclosure and on a level parallel to the bottom of such primary enclosure at a point which approximates half the distance between the top and bottom of such primary enclosure.

[43 FR 56217, Dec. 1, 1978. Redesignated at 44 FR 36874, July 22, 1979]

§ 3.142 - Handling.

(a) Carriers and intermediate handlers shall move live animals from the animal holding area of the terminal facility to the primary conveyance and from the primary conveyance to the animal holding area of the terminal facility as expeditiously as possible. Carriers and intermediate handlers holding any live animal in an animal holding area of a terminal facility or in transporting any live animal from the animal holding area of the terminal facility to the primary conveyance and from the primary conveyance to the animal holding area of the terminal facility, including loading and unloading procedures, shall provide the following:

(1) *Shelter from sunlight.* When sunlight is likely to cause overheating or discomfort, sufficient shade shall be provided to protect the live animals from the direct rays of the sun and such live animals shall not be subjected to surrounding air temperatures which exceed 29.5 °C. (85 °F), and which shall be measured and read in the manner prescribed in § 3.141 of this part, for a period of more than 45 minutes.

(2) *Shelter from rain or snow.* Live animals shall be provided protection to allow them to remain dry during rain or snow.

(3) *Shelter from cold weather.* Transporting devices shall be covered to provide protection for live animals when the outdoor air temperature falls below 10 °C. (50 °F) and such live animals shall not be subjected to surrounding air temperatures which fall below 7.2 °C. (45 °F.), and which shall be measured and read in the manner prescribed in § 3.141 of this part, for a period of more than 45 minutes unless such animals are accompanied by a certificate of acclimation to lower temperatures as prescribed in § 3.136(c).

(b) Care shall be exercised to avoid handling of the primary enclosure in such a manner that may cause physical or emotional trauma to the live animal contained therein.

PART 3
Subpart F

(c) Primary enclosures used to transport any live animal shall not be tossed, dropped, or needlessly tilted and shall not be stacked in a manner which may reasonably be expected to result in their falling.

[43 FR 21167, May 16, 1978, as amended at 43 FR 56217, Dec. 1, 1978. Redesignated at 44 FR 36874, July 22, 1979]

PART 3
Subpart F

PART 4 – RULES OF PRACTICE GOVERNING PROCEEDINGS UNDER THE ANIMAL WELFARE ACT

Subpart A – General

§ 4.1 Scope and applicability of rules of practice.

Subpart B – Supplemental Rules of Practice

§ 4.10 Summary action.
§ 4.11 Stipulations.

Authority: 7 U.S.C. 2149 and 2151; 7 CFR 2.22, 2.80, and 371.7.
Source: 42 FR 10959, Feb. 25, 1977, unless otherwise noted.

Subpart A – General

§ 4.1 - Scope and applicability of rules of practice.
The Uniform Rules of Practice for the Department of Agriculture promulgated in subpart H of part 1, subtitle A, title 7, Code of Federal Regulations, are the Rules of Practice applicable to adjudicatory, administrative proceedings under section 19 of the Animal Welfare Act (7 U.S.C. 2149). In addition, the Supplemental Rules of Practice set forth in subpart B of this part shall be applicable to such proceedings.

Subpart B – Supplemental Rules of Practice

§ 4.10 - Summary action.
    (a) In any situation where the Administrator has reason to believe that any person licensed under the Act has violated or is violating any provision of the Act, or the regulations or standards issued thereunder, and he deems it warranted under the circumstances, the Administrator may suspend such person's license temporarily, for a period not to exceed 21 days, effective, except as provided in § 4.10(b), upon written notification given to such person of the suspension of his license pursuant to § 1.147(b) of the Uniform Rules of Practice (7 CFR 1.147(b)).

    (b) In any case of actual or threatened physical harm to animals in violation of the Act, or the regulations or standards issued thereunder, by a person licensed under the Act, the Administrator may suspend such person's license temporarily, for a period not to exceed 21 days, effective upon oral

PART 4

or written notification, whichever is earlier. In the event of oral notification, a written confirmation thereof shall be given to such person pursuant to § 1.147(b) of the Uniform Rules of Practice (7 CFR 1.147(b)) as promptly as circumstances permit.

(c) The temporary suspension of a license shall be in addition to any sanction which may be imposed against said person by the Secretary pursuant to the Act after notice and opportunity for hearing.

§ 4.11 - Stipulations.

(a) At any time prior to the issuance of a complaint seeking a civil penalty under the Act, the Administrator, in his discretion, may enter into a stipulation with any person in which:

(1) The Administrator gives notice of an apparent violation of the Act, or the regulations or standards issued thereunder, by such person and affords such person an opportunity for a hearing regarding the matter as provided by the Act;

(2) Such person expressly waives hearing and agrees to pay a specified penalty within a designated time; and

(3) The Administrator agrees to accept the specified penalty in settlement of the particular matter involved if it is paid within the designated time.

(b) If the specified penalty is not paid within the time designated in such a stipulation, the amount of the stipulated penalty shall not be relevant in any respect to the penalty which may be assessed after issuance of a complaint.

INDEX

INDEX

INDEX

245

INDEX

INDEX

INDEX

INDEX

249

INDEX

INDEX

 United States
Department of
Agriculture

Made in the USA
Monee, IL
24 April 2024

57470866R00144